realistic cost estimating for manufacturing

IVAN R. VERNON

Editor

PUBLISHED BY:

SOCIETY OF MANUFACTURING ENGINEERS

Dearborn, Michigan
1968

**Realistic Cost Estimating
for Manufacturing**

Library of Congress Catalog Card Number: 68-8828
SBN: 87263-012-9

FIRST EDITION ■ MANUFACTURING DATA SERIES
MANUFACTURED IN THE UNITED STATES OF AMERICA

PREFACE

In the past, manufacturers frequently initiated production of many items without a detailed examination of material, labor, and tooling costs. With the sharper competition and increased costs existing today, this course is no longer economically feasible.

Careful product cost estimates have become an accepted and essential element of manufacturing planning. The "guesstimate" and the estimator's "little black book" receive little credence in a day when a cost differential of $.006 on a $.06 item can provide the basis for a make-or-buy decision, when a cost rise or reduction of $.01 on a $1.00 item can spell the difference between profit or loss. Detailed cost analyses and reliable standard data have taken their place.

This book aims to assist estimators in performing their important function in the manufacturing organization. Chapter 1, "The Estimating Function," defines the customary terminology, explains the manufacturing cost structure, discusses the purpose of estimating, and describes the various types of estimates. Chapter 2, "Organization and Staffing for Estimating," analyzes the role of the estimating function in the corporate structure, discusses departmental organization, sets forth the necessary qualifications of an estimator, examines appropriate training programs, offers suggestions regarding personnel policies for estimating departments and discusses computerized estimating.

The balance of the book is devoted to the actual estimating procedure. Chapter 3, "Cost Estimating Controls," describes the desirable administrative controls, reviews the method by which cost requests are initiated, scrutinizes the various estimating methods, examines the different methods of controlling estimate deviations, and concludes with a section setting forth the "do's and don't's" of estimating. Chapter 4, "Estimating Procedures," explains in detail the steps necessary to estimate the cost of a manufactured item. Chapter 5, "Cost Estimating Examples," provides examples of completed cost estimates on several manufactured products that can be used as a guide for making nearly any type of estimate.

Illustrations are used liberally throughout the book to demonstrate specific points under discussion. For instance, examples of estimating forms presently being used by various manufacturers are included to help estimators devise suitable forms to fit their own needs.

SME wishes to thank the following individuals for reviewing the manuscript and offering many helpful suggestions: William J. Van Dyke, Cost Estimating Manager, Rausch Manufacturing Company, St. Paul, Minnesota; Desmond Douglas, Key Part Control Analyst, Ford Motor Company, Dearborn, Michigan; and Joseph Seman, Cost Estimating Chief, Norair Division, Northrop Corporation, Hawthorne, California. Special thanks are due Maurice Collin, Coordinator of Product Tool Studies, Chevrolet Motor Division, General Motors Corporation, who advised the editor during the preparatory stages of the manuscript as well as serving as a reviewer.

Dearborn, Michigan   IVAN R. VERNON
November, 1968

TABLE OF CONTENTS

CHAPTER 1   **THE ESTIMATING FUNCTION** **1**
Explanation of Terms **1**
Manufacturing Costs **2**
Purpose of Estimating **6**
Types of Estimates **7**

CHAPTER 2   **ORGANIZATION AND STAFFING FOR ESTIMATING** **9**
The Estimating Function in the Corporate Structure............ **10**
Departmental Organization **12**
Qualifications of an Estimator **13**
Development of an Estimator **15**
Personnel Policies for Estimating................................ **15**
Computer Applications to Cost Estimating........................ **15**

CHAPTER 3   **COST ESTIMATING CONTROLS** **19**
Administrative Controls **19**
Initiating Cost Requests **21**
Estimating Methods **26**
Controlling the Cost Estimate **28**
Do's and Don't's of Cost Estimating **33**

CHAPTER 4   **ESTIMATING PROCEDURES** **35**
Estimate Analysis **36**
Part Analysis **37**
Preliminary Manufacturing Plan **38**
Facilities Estimating **39**
Direct Material Cost **41**
Tooling Costs **44**
Manufacturing Time Periods **45**
Direct Labor Costs **47**
Factory Burden **47**
Total Manufacturing Cost **49**
Selling Price **49**

CHAPTER 5   **COST ESTIMATING EXAMPLES** **51**
Estimating Die Casting Machining Costs........................ **51**

Estimating Machining Costs for an Aluminum
    Forging.. 58
Screw Machine Cost Estimating......................... 64
Estimating Sand Casting Costs 77
Estimating Welding Costs................................. 89
Estimating Forged Parts..................................100
Estimating Metal Stamping Costs 109
Estimating Plastic Parts...................................113
Estimating Tumbling and Vibratory Finishing
    Costs..118

BIBLIOGRAPHY...123

INDEX 125

WORKBOOK 133

LIST OF TABLES

Table IV-1	Cost Estimating Data	37
Table IV-2	Machinery Installation Costs	40
Table IV-3	Cost Estimate for Automobile Bumper	48
Table V-1	Standard Data for Time Study	54
Table V-2	Standard Data for Drilling, Reaming, and Tapping Aluminum	56
Table V-3	Data for Milling Machine	56
Table V-4	Summary of In-Plant Manufacturing Operations and Their Costs	57
Table V-5	Derivation of Selling Price for Aluminum Forgings in Lots of 100	63
Table V-6	Typical Labor Rates per 100 Minutes	63
Table V-7	Cutting Speed and Machinability Rating for Steels	68
Table V-8	Spindle Speeds and Revolutions per Piece for Laying Out Cams	70
Table V-9	Approximate Cutting Speeds and Feeds for Standard Tools on a Single-Spindle Automatic Screw Machine	71
Table V-10	Angles and Thicknesses for Circular Cutoff Tools	72
Table V-11	Machine Spindle Revolutions for Part in Fig. 5-7a	73
Table V-12	Machine Spindle Revolutions for Part in Fig. 5-7b	75
Table V-13	Hundredths Required to Index for Throws from Full-Height Cam	76
Table V-14	Clearance in Hundredths between Turret Tools and Cross-Slide Circular Tools	77
Table V-15	Record of Weights of Gray Iron Castings	79
Table V-16	Record of Weights of Malleable Iron Castings	80
Table V-17	Calculation for Weight of Water Outlet Casting	85
Table V-18	Data for Fillet and 45-Deg. Bevel Welds	97
Table V-19	Data for Spot Welding	99
Table V-20	Density of Metals	102

Table V-21   Approximate Flash Thickness and Width on Forgings **103**

Table V-22   Typical Forge Shop Estimator's Chart.........................**105**

Table V-23   Calculation for Weight of Gear Blank in Fig.
5–23 ....**108**

Table V-24   Media Attrition Data for a Tumbling Barrel
Finishing Machine..**122**

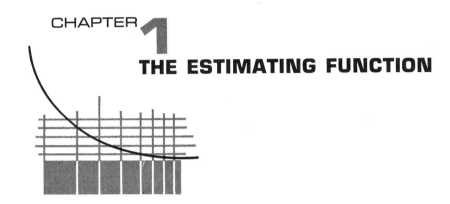

CHAPTER 1

THE ESTIMATING FUNCTION

Cost estimating is defined as an attempt to predict the expenses that must be incurred to manufacture a product. The total estimated cost figure consists of direct labor, raw materials, tooling, and fixed and variable factory burden rates. All these items combine to form the *product cost estimate,* to which the sales and/or accounting departments add a percentage for profit.

Estimating is a staff function. The estimating department uses its expertise in product costing to provide management and operating departments with cost data for the purposes of bidding for new jobs, contracts, etc., evaluating proposed products, comparing alternative product designs, and deciding make-or-buy situations. The estimating department is only one of several functions responsible for manufacturing costs. Others include accounting, cost control, value analysis, and value engineering. Each of these functions and their relationships to cost estimating are described later in this chapter.

EXPLANATION OF TERMS

Estimates used for product costing may be classified in several ways. The term *cost estimate* generally refers to any estimate of the costs involved in the manufacture of a piece part, subassembly, or whole product. Occasionally the term *part estimate* is used to refer to the estimate of the cost of manufacturing a product subassembly or piece part. Also, the term *product estimate* is sometimes used to emphasize the fact that an entire product rather than a subassembly or piece part is being estimated.

The term *piece part* refers to the smallest component of a product or product subassembly. A *subassembly,* composed of two or more piece parts, can be defined as a subsystem. An automobile exhaust system, for example, is a subassembly. The tailpipe and the clamps and bolts attaching the tailpipe to the automobile are piece parts. A product usually consists of a number of piece parts and one or more subassemblies.

1

The cost estimate is composed of various cost categories. When necessary to distinguish between major cost components, the terms *material cost estimate, labor cost estimate,* and *tooling estimate* are useful. In recent years, tooling costs have increased to such an extent that most large companies assign tooling costs to individual parts rather than including them in factory burden.

Estimating is necessary in areas other than product costing. The terms *facilities cost estimate, project cost estimate,* and *construction cost estimate* suggest the important role estimating plays in long-range manufacturing planning, though the responsibility for such planning is usually held by a facilities planning group or some other planning group rather than being delegated to the cost estimating department.

MANUFACTURING COSTS

An acquaintance with the manufacturing cost structure is essential to an understanding of the estimating function.

Types of Costs

Manufacturing costs are classified one way as *direct* and *indirect.* A second classification includes *actual* and *standard* costs.

Direct Costs. Direct costs are those that can be traced directly to a specific piece part, subassembly, or product. Any other cost is an indirect cost, one that cannot be identified with the manufacture of a specific product.

Direct costs include the cost of tools designed or used specifically for a particular part, the cost of the material from which the part is fabricated, and the cost of the labor used to make the part.

The cost of the metal from which an automobile fender is formed would always be a direct cost. The wages paid the stamping machine operator would likewise be classified as a direct cost. Secretarial and accounting expenses would always be indirect costs. But between these distinct areas lies a gray region. What about the cost of the gas required to heat a baking kiln for curing painted parts? It may be impracticable to assign this cost proportionately to each part passing through the kiln.

Thus the estimator or accountant asks two questions before classifying a cost as direct or indirect: How directly is the cost related to the manufacture of a specific product? How practicable is it to relate the cost to a particular item? Relatively small costs may be classified as indirect simply because it is not worthwhile to break the costs out to specific products.

Indirect Costs. Indirect costs cover those items necessary to operate the manufacturing plant but not traceable directly to one specific product. The costs of janitorial service, forklift operators, machine maintenance, utilities, and certain nonassignable materials and tooling are all indirect.

Indirect costs are also broken down further into factory burden, and general and administrative overhead. As a rule, cost estimators are concerned with factory burden but not with general and administrative expenses. Usually the sales or accounting department, in setting the product selling price, adds an amount for general and administrative expenses to the cost estimate. In some companies, however, the estimating department may assign general and administrative costs, using information supplied by the accounting department.

Factory Burden. Factory burden includes all costs associated directly with the operation of the plant but not directly attributable to a particular product. Utilities for lights and heating, the labor and material for factory housekeeping, durable tooling usable for many different products, and such indirect materials as cutting and grinding fluids are examples of burden items.

General and Administrative Overhead. General and administrative costs are those necessary to maintain the firm in operation but not directly applicable to the production function. Executive salaries, long-range research and development expenses, and public relations costs are examples.

Standard Versus Actual Costs. Manufacturing costs may also be classified as standard or actual. A standard cost is an ideal cost or a predetermined cost. Working with the manufacturing department, cost accounting develops figures reflecting what a part or product *should* cost. When the part is completed or the product is finished, accounting determines what the *actual* costs were.

The variances between actual costs and standard costs provide management with a tool to evaluate the effectiveness of first-line supervisors. Variances below the standard may indicate that a foreman or department supervisor has found ways to reduce materials wastage, cut overtime, or decrease tool breakage and wear. Costs higher than the standard may be a sign of high wastage, poor workmanship resulting in a high part rejection rate, or abnormally high labor costs due to lax supervision or excessive overtime.

On the other hand, variances above standard costs may reflect, not ineffective management but, rather, may indicate an uncontrollable cost increase. In this case, variances serve as an effective reporting system. By readily identifying costs that are out of line, management can take appropriate action. One way would be to adjust standard costs so that future estimates used for bidding are more realistic.

Costs in Relation to Product Volume

Costs are also categorized according to their behavior in relation to product volume. Fixed, variable, and step-variable cost patterns are shown in Fig. 1–1. The definitions of fixed, variable, and step-variable costs assume the existence of upper and lower limits on production quantities. Only within this relevant range do costs behave according to their definitions. Note the cost behavior patterns below 50 units and above 250 units in Fig. 1–1.

Fixed costs remain the same regardless of volume. Specific examples include executive salaries, secretarial services, facilities, durable tooling, and security services. Durable tooling is classified as a fixed cost because a certain level of tooling is required to produce one unit or many units. Adequate plant security probably demands at least one night watchman, regardless of production levels.

Variable costs rise in direct proportion to the number of units produced. Direct costs such as material, labor, and perishable tooling generally follow a variable curve such as the one depicted in Fig. 1–1. Note that this curve climbs less steeply outside the relevant range. As larger volumes are produced, economies on larger material and tooling purchases are possible, thus lowering costs per unit.

A *step-variable cost* is one that remains the same over a given number of production units but jumps sharply to new plateaus at certain incremental changes in volume. The cost of a machine capable of producing 50 units, for example,

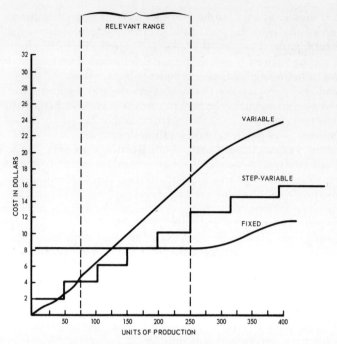

Fig. 1–1.   Cost-volume behavior patterns.

would be the same whether one or 50 units were actually produced. At 51 units, however, a new machine would need to be purchased, thus doubling the cost of machinery.

Cost Responsibility

The ultimate responsibility for all aspects of a manufacturing firm's costing, pricing, and financial system rests with management. Management retains direct control in such areas as long-range financing, capital investment, and major product development (R&D). This control is exercised with the assistance and advice of staff personnel such as attorneys, cost accountants, research and development engineers, etc. In other areas, management relinquishes operating responsibility to subordinate departments. Operating under broad guidelines established by management, for example, the cost estimating department or the purchasing department may be responsible for certain make-or-buy decisions.

Distinctions should be drawn between cost estimating and those functions also principally concerned with costs—accounting and cost-reduction activities. Cost estimating differs from accounting in that it is a preproduction function; it is thus a forecast, an opinion, an appraisal, an educated guess, or a scientific analysis of all the factors entering into the manufacture of a proposed product. Accounting, on the other hand, is historical in nature. It involves the recording, analysis, and interpretation of events, or costs, that have already transpired.

Cost estimating gives management information to help make product development decisions, while accounting provides information to assist in evaluating decisions made earlier.

Accounting Department. The accounting department records, monitors, and reports the results of the company's external financial transactions, traditionally called financial accounting. It involves posting new information to a book of original entry, transferring such information to classified accounts, and preparing various reports based on these accounts for the use of management and sometimes the public. Reports for management contrast and compare the results of various past activities to assist in making intelligent, informed decisions regarding future courses of action. The profit and loss statement, balance sheet, and cash flow statement, combined, present a total picture of financial transactions and resources of the firm for the benefit of investors, bankers, customers, the U.S. Internal Revenue Service, etc.

Accounting also records, monitors, and reports the results of financial transactions within the firm. In addition to routine payroll activities, it develops standard costs and reports variances, compiles reports on inventory levels and costs, and computes depreciation amounts on assets held by the company. This responsibility is known as cost accounting. It involves gathering accurate cost data for various product inputs and establishing cost guidelines for the managers of production departments.

Management evaluates departmental efficiency on the basis of variances from these guidelines or standard costs. Through analysis of such cost figures, management may also adjust internal operating procedures so as to improve efficiency. Analysis, for example, may suggest early replacement of a machine that has become too expensive to use. Or management may decide to change the product mix in order to take advantage of differences in the prices of material or labor. With union agreement, for example, the company might effect a savings by assigning one well-paid skilled worker to a task previously performed by two unskilled workers.

While financial accounting reports the results of relationships with parties in the exterior environment, cost accounting provides a means of understanding internal activities and endeavoring to control them to keep the firm competitive.

Cost Reduction. Cost control, value analysis, and value engineering are three related concepts, each concerned broadly with cost reduction, but differing in their orientation. Most large manufacturing companies have accepted one of these concepts and established a department to carry out the necessary activities.

Cost Control. Cost control is primarily concerned with reducing the cost of manufacturing an item while retaining the desired quality and performance. Lowering the scrap rate, increasing tool life, and eliminating unnecessary processing steps are all proper responsibilities of cost control.

Value Analysis. The purpose of value analysis is to optimize value added to a product by reducing manufacturing costs while maintaining or enhancing product functionality. The value analysis department analyzes a product according to its attributes, at the same time keeping in mind the capabilities of the plant and the cost of securing materials, parts, subassemblies, and production services from outside sources. The object of this analysis is to examine all available alternatives in order to determine which will produce the best value; which design, for example, will give the most reliable performance and have the greatest sales appeal for the least cost. To accomplish this, value analysis considers factors related to internal as well as external environments.

Value Engineering. Value engineering is a concept which goes beyond cost control and value analysis and represents an attempt to combine these two functions on a systems basis. Value engineering examines all of the manufacturing activities of the firm, seeking to optimize both long-range and short-range product profitability. It not only attempts to optimize value added, but also analyzes necessary production services in an attempt to optimize them in such a manner as to maximize product quality while minimizing costs. Effective value engineering requires up-to-date knowledge of production processes, materials, and tooling. Viewed conceptually, it is an attempt to engineer production economy and quality into a product. To achieve these purposes, value engineering should begin in the product design stage.

Cost Estimating. Where does cost estimating fit into the manufacturing cost structure? As defined earlier, cost estimating is an attempt to predict the costs that must be incurred to manufacture a product. While accounting deals generally with historical data, estimating attempts to peer into the future. While value engineering takes a systems approach to manufacturing costs, estimating scrutinizes the individual elements of product cost. While the estimating department searches out and applies cost information to predict the cost of a given product, value analysis and cost control must discern apparent economies.

From an organizational standpoint, cost-reduction activities work closely with the estimating department to a greater or lesser degree. Before instituting a change in the manufacturing plan, for example, a value analysis department might forward the proposal to the estimating department for detailed evaluation. On the other hand, the relationship between the accounting and estimating departments is mainly one of exchanging information.

PURPOSE OF ESTIMATING

The cost estimating function is important to any manufacturing organization. A carefully prepared estimate is essential in deciding whether to begin manufacturing a product, for instance. In the overall planning process, the estimating phase is the fundamental step in determining product costs, tooling costs, and lead time. Managers also need cost estimates for make-or-buy decisions, in bidding on contracts, and in evaluating the products of competitors or vendors.

The product estimates prepared by estimating for management are used to make decisions about both present and future courses of action. In addition, operating departments use estimates as guides to the relative cost of equipment, tools, and services necessary to produce an item. The estimating department, for example, supplies cost data to other phases of the organization such as process planning, tool design, materials handling, and plant layout for their use in planning for new products.

Estimating itself adds no value to products; it does not affect the final costs except as it provides temporary standards. It is, however, a valuable tool for evaluating and comparing manufacturing alternatives and materials, and developing design proposals. Since value added to the product cannot be measured quantitatively, estimating methods should be selected and organized to produce optimum return from allocated resources. Little is gained in preparing elaborate estimates if simpler ones will furnish comparable data.

A carefully executed cost estimate based on a well-defined product promises a high degree of probability that the activity planned, if undertaken, will work out as calculated. Good cost estimates may be used to:

1) Establish the bid price of a product for a quotation or contract
2) Verify quotations submitted by vendors
3) Ascertain whether a proposed product can be manufactured and marketed profitably
4) Provide data for make-or-buy decisions
5) Help determine the most economical method, process, or material for manufacturing a product
6) Provide a temporary standard for production efficiency and guide operating costs at the beginning of a project
7) Help in evaluating design proposals.

TYPES OF ESTIMATES

Cost estimates are one of two types: preliminary or final. Preliminary estimates are used for product design guidance and to provide cost information to management in the early stages of product planning.

Final cost estimates are prepared once product specifications are finalized and before parts are actually produced. Final estimates are also made for major projects other than products, such as R&D investigations. In addition, tooling, equipment, and facilities should all receive careful cost evaluation by means of detailed estimating in order to determine profitability.

Preliminary Estimate

Estimating departments usually make preliminary cost estimates for new parts or products before designs and plans are complete. Process engineering and design engineering departments use this type of estimate to compare alternative designs or manufacturing methods to determine the most economical processes or designs.

If the design engineering department is considering a product change from a metal tub to a blow-molded plastic tub in the company's line of washing machines, for example, the engineers might request a rough estimate of the cost of the new design proposal. The estimating department then prepares a preliminary estimate giving the approximate cost of material, labor, tools, and equipment for the new design. Using this information, management makes a decision by studying the cost differential between the new and old parts, as well as factors affecting quality. After the design decision is made, final production plans are prepared, and eventually a final cost estimate will be needed. (See Fig. 3-3.)

It is easier for the estimator, individual manufacturing departments, or an estimating conference (see Chapter 3) to make preliminary estimates when drawings or layouts are available. However, if drawings are not available, engineers familiar with the project brief the individual estimator or estimating group on the available data, sometimes supplying what estimators call "matchbook cover specifications." From this briefing, a preliminary estimate of the cost of manufacturing the product can be made by the estimator or other persons responsible.

Fig. 1–2 shows a typical form used for preliminary and final cost estimates.

PART COST ESTIMATE								
PART NO.	PART NAME	STANDARD TIME	LABOR	VAR.	FIXED	TOTAL	TOOLS	REMARKS

PART NAME _____ PART NO. _____
MATERIAL SPEC. _____ QUANTITY PER JOB _____ YEAR _____ MODEL _____
ROUGH MATERIAL SIZE _____ ROUGH WEIGHT _____ #@ ____/#= ____ MATERIAL COST PER PIECE

PER PIECE / PER. VEH.
MAT'L
LABOR
BURDEN
TOTAL

REQUESTED BY: _____
FROM B/P DATED: _____
ESTIMATED BY: _____
DATE _____ SHEET ____ OF ____

Fig. 1–2.   Form used by one company for preliminary and final cost estimates.

Final Estimate

A final or detailed cost estimate, based on fully developed production plans and designs, is obviously the most accurate. The final estimate serves as a performance standard or may be used to evaluate vendor prices or to assist in make-or-buy decisions. A detailed estimate is made for every component, assembly, and subassembly and must reflect the cost of every tool or production part with reasonable accuracy.

Complete, reliable data cannot be developed without the expenditure of sufficient time. The amount of time that should be spent on a particular estimate is frequently a function of the amount of money involved. For instance, if 1,000,000 units costing $1.00 each are to be produced, the estimate should not be rushed through as fast as the estimate for 10,000 units costing $.50 each. Often the work involved to obtain the most precise type of estimate is not justified. Ascertaining precise cost data for use in tandem with data developed as a rough estimate is uneconomical. In computing return on investment, for example, it is a waste of time to compile precise figures for capital costs for use with estimated figures representing sales volume and price structure. But when precise costs are available (e.g., for capital investment) they may be used to compute exact product profit margins required to make the investment feasible.

CHAPTER 2

ORGANIZATION AND STAFFING FOR ESTIMATING

While the estimating function is essential to any manufacturing organization, the importance of the estimating role, and consequently, where it is located in relation to other departments in the organization, varies from company to company. The responsibilities delegated to the estimating department of a particular company are a primary determining factor in departmental organization. The interest and capabilities of the individual estimators also effect the departmental setup. Product complexity and other factors should be considered, but the overriding determinants are efficiency and speed of operation for the department as a whole.

The estimating staff requires special attention. Candidates should be evaluated in terms of their experience, learning capabilities, and personality traits because only particular types of individuals can cope with the uncertainty, heavy responsibility, and relatively unstructured work situation inherent in estimating.

Basic manufacturing experience is desirable, and time-and-motion study is one of the best backgrounds. Candidates selected must be capable of being trained. Most companies use on-the-job training for new estimating personnel, but a few companies have set up more formal courses. Various training aids, such as programmed learning courses, are also available.

Once a trained estimating staff has been assembled, management should attempt to reduce turnover so that the investment in training and the on-the-job experience gained by the estimator accrue to the benefit of the company. Estimators respond favorably to the same professional treatment accorded engineers and other highly trained manufacturing personnel. Supplying professional needs for responsibility, authority to do their jobs, recognition for good performance, and the best possible working environment to ensure technical competency is a major duty of management.

9

THE ESTIMATING FUNCTION IN THE CORPORATE STRUCTURE

Cost estimating in most manufacturing organizations is performed by an experienced person called by various titles including *estimator, purchase cost analyst,* and *manufacturing cost analyst.* The estimating function is generally placed under the general supervision of the manufacturing engineering or accounting departments. The administrative supervisor of the department's activities may be the plant manager, chief manufacturing engineer, or comptroller.

In some companies, the estimating function is responsible to the sales or marketing department. Generally this is not desirable because sales departments, utilizing modern marketing concepts, tend to be heavily customer oriented, and salesmen may occasionally attempt to increase sales volume at the expense of the firm's profit margin. Situations occur, for example, when a sales department may seek to increase sales in one product line by offering the customer a bargain in another product line. In addition, sales or marketing may have special reasons for wanting to develop products for particular customers, regardless of manufacturing costs. Although the marketing tactic of proffering "price breaks" is valid, it is essential that both sales and manufacturing consistently recognize and apply accurate product costs in plotting overall corporate strategy in the areas of marketing and production.

Fig. 2–1 shows a typical organization chart for medium to large single-plant companies. In the organization this chart portrays, the cost estimating function reports to the manufacturing engineering department which, in turn, is supervised by the manufacturing manager. The departments with which the estimating department works most closely are indicated by the heavier lines.

Cost Estimating Systems

Two types of cost estimating systems are commonly found in industry. One, an integrated system, utilizes a department which makes the complete estimate based on data in their files or data obtained from various sources for a particular estimate. The second is a departmentalized system in which cost data for individual manufacturing processes are added to the total estimate by the department responsible for conducting that activity. For example, methods or time study engineering applies standard time data to the manufacturing plan or operation lineup furnished by the process planner, accounting extends the time data by applying labor rates and burden factors, and tool design adds estimated tooling costs.

Since it takes two or three years to develop a self-sustaining estimating department, the second type of cost estimating system is recommended for new companies or older companies just starting an estimating function. However, this estimating system has several disadvantages. First, unless all of the participating departments are fully staffed and experienced in this type of procedure, they may regard estimating as an unnecessary expenditure of time and effort and thus give little attention to the estimating function. Second, the normal human tendency to overprotect one's department by estimating costs at easily achievable levels can result in the loss of bids. And third, the time necessary to process this type of estimate may prevent the meeting of deadlines.

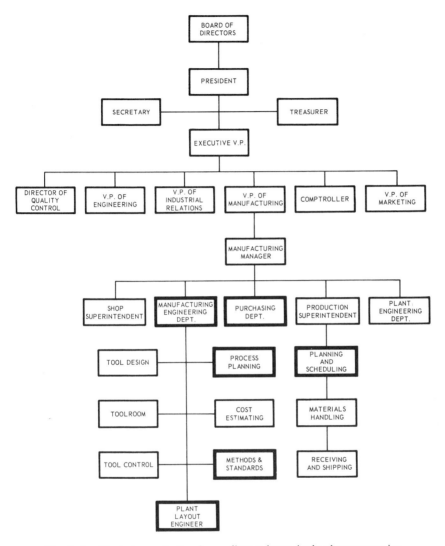

Fig. 2–1.   Typical organization for medium to large single-plant companies.

Top management must decide what kind of estimating organization is best for its particular situation. The advantages and disadvantages of each must be carefully weighed. Among the many questions that must be answered to determine whether a company should have an integrated or departmentalized estimating function are:

1) What is the normal lead time required to prepare an estimate? If estimates for complex assemblies must be completed in less than one week, it may be impossible to coordinate all participating departments. When the services of specialists are required, lead time is particularly critical.

2) How many people are needed for an integrated estimating department? This question can best be answered by a question: What type and size of manufacture is involved? The small job shop will usually require only the services of one highly capable estimator. A large job-shop manufacturer, on the other hand, producing many complex products in small production lots will require many times the number of estimators needed by a continuous processing industry with comparable sales figures.

3) Is it possible to draw personnel from within the organization to establish an estimating department? Not only must they be potential estimators, but can they be replaced in their existing capacities?

4) Can a company sustain the cost of a full-time estimating department? Studies invariably indicate that a company cannot afford to be without one. It is highly improbable that any company will survive the ordeals of competition without the services of a capable estimator.

5) How will the estimating function be maintained during the one, two, or three years required to develop an independent estimating department? Some interim measure, such as on-the-job training coordinated with departments previously supplying estimating information, may be arranged. Process planning, methods engineering, purchasing, and other functions will, from time to time, be called upon for assistance with knotty problems, even by the ablest of estimating departments.

Estimating and the Product Development Cycle

Top management must also determine the stage of the product development cycle at which the estimating department should become involved. This differs according to whether the company develops most of its business by bidding, or whether it has an established product line marketed to a relatively stable group of customers.

In the first instance, rather accurate cost estimates must be developed before a bid is submitted. This means that the estimating department will frequently become concerned with material selection, product configuration, and process planning earlier than any other department in the organization. Firms with a heavy volume of bidding sometimes have two separate estimating departments. One is responsible for preparing contract bid estimates, the other for developing piece part or product estimates on accepted contracts.

Manufacturing organizations developing products to market to a wider segment of industrial customers and wholesale distributors can afford more time for developing estimates. In this case, the estimating department may not become involved until after a great deal of the manufacturing planning is completed.

DEPARTMENTAL ORGANIZATION

Similar to manufacturing functions, individual tasks in the estimating department should be carefully delineated, responsibilities plainly stated, and lines of authority clearly defined. Organizational ambiguities are as harmful to successful estimating as to any other function.

Two basic methods of organization exist: product or specialty oriented and nonspecialized. Company size and type of product will determine the best method for a particular company.

Product or Specialty Oriented

Large, diversified manufacturers sometimes organize their estimating departments on a product basis so that an individual estimator always works on related items. Or, estimating departments in plants with forging, stamping, casting, and plastic capabilities may have specialists in each of these areas. In this manner the estimator can gain continual related experience and compile a comprehensive data file.

In a large automobile manufacturing plant, for example, an estimator may be a specialist in fuel and exhaust components, body parts, or engines. But he may nevertheless handle a wide variety of assignments. He may develop cost data for engineering changes affecting current or future models, or he may be assigned a competitor's or vendor's part for competitive analysis.

Nonspecialized

Estimating departments not structured by specialty usually operate on a first-come, first-served basis. Cost estimate requests received in the department are assigned to the estimator who is available at the moment. The advocates of this type of organization claim that it is more flexible over a long period of time.

The chief advantage of nonspecialization is that individual estimators gain the experience necessary to provide estimates on the company's entire product line. Thus a nonspecialized department is more flexible but not necessarily more efficient. Some inefficiency may be the price that must be paid for flexibility.

The question of specialization versus nonspecialization may answer itself for any individual company. For instance, companies manufacturing a homogenous line of fairly simple products may not have sufficient product variation to make specialization possible. Also, small estimating departments may have no alternative to the first-come, first-served method. Estimators serving job shops that manufacture a wide range of products in small lot sizes have little opportunity to specialize.

Estimating Supervisor

An experienced estimator usually makes a good supervisor for the estimating department. His duties include general administration, work assignment, and estimate checking. Whether supervisory duties will occupy the department head's full time, or whether he can also perform estimating depends upon his workload and upon company policies. In some companies the estimating supervisor attends cost meetings, coordinating estimating activities with other departmental activities.

QUALIFICATIONS OF AN ESTIMATOR

What are the attributes to look for in the selection of a candidate for estimating? Basically an individual should have an analytical mind and a scientific approach. He must be able to observe, generalize from observed facts, and check generalizations by further observations, past experience, accumulated factual data, and accepted standards.

Cost estimators should possess the following traits and qualifications:

1) Education and/or experience in time-and-motion study and methods analysis (probably the most important qualification)
2) The ability to observe and retain data accurately
3) The ability to reason scientifically or approach a problem systematically
4) The analytical mind of an engineer
5) Knowledge of general accounting through manufacturing costs
6) Tool "know-how," including toolroom and tool troubleshooting experience
7) Process planning knowledge and/or experience
8) General knowledge of materials and their properties
9) Knowledge of the capabilities and limitations of the machines and tools in his plant
10) The ability to work quickly and accurately
11) The initiative or drive to investigate and keep abreast of the latest advances in processes and materials in his manufacturing field, without which his skills will quickly become obsolete.

Knowledge of manufacturing processes in general and the labor and methodology required by each operation is extremely important both in estimating in-house jobs and in estimating the cost of a competitor's or vendor's product. The greater the estimator's understanding of manufacturing processes, procedures, and techniques, the more valuable he is.

The most important qualifications of an estimator are developed only through long years of experience. Because the availability of such individuals is limited, companies often must recruit employees capable of being trained, rather than fully trained estimators.

An estimator must be able to recognize his inherent bias. Checking the accuracy of previous estimates against presently known costs will serve to reorient the estimator and help him reverse any trend toward inaccuracy. From his periodic self-examination, the estimator can adjust his estimates up or down according to his past experience. The estimator who unconcernedly continues to prepare estimated costs without correcting for his degree of error could conceivably bankrupt his company through loss of sales or loss of profit.

A final factor to consider in determining the qualifications of a potential estimator is his ability to visualize precise details from poorly developed specifications. Customers, engineers, and sales department personnel frequently jot down a few words and draw a few lines on the back of an envelope when requesting "ballpark" figures or quick cost information on a new product. In this situation the estimator must be familiar with the product line, and be enough of an engineer to ask judicious questions and fill in the specifications sufficiently in order to develop an estimate. He must understand the quality, performance, and reliability intended for the product, and he must know what finishes, tolerances, and inspections will be necessary to achieve them. The estimator must also obtain confirmation of his assumptions on those points that are not clear to him from a knowledgeable source.

DEVELOPMENT OF AN ESTIMATOR

An estimator is not trained quickly but evolves slowly by exposure to many facets of manufacturing. A man with a diversified background in processing and time-and-motion study is the most easily trained. The remaining traits may be gained by formal education and/or in-plant training.

College graduates with degrees in industrial, mechanical, manufacturing, or tool engineering combined with manufacturing experience have proved to be the most readily adaptable to cost estimating. Supplemental courses in mathematics, accounting, and time-and-motion study, along with a specialized course in the field in which the estimator plans to work, are beneficial.

An in-plant training program would round out the estimator's understanding of the particular company's policies, procedures, and products. The length and amount of training are governed by the individual and his responsibilities. In normal fabricating operations, a designated period of time spent in process planning, time-and-motion study, and tool planning groups comprise the training program, but on-the-job training in the estimating department is especially valuable. Challenging assignments, combined with a training director and schedule, serve to motivate the trainee.

Many companies have begun to recognize the value of formal orientation courses for new estimating personnel. Programmed learning courses, commercially available textbooks, and other instructional materials are often used for such courses. Larger companies sometimes prepare their own course outlines, text materials, and provide a qualified instructor.

In addition, engineering colleges and professional engineering societies offer seminars and workshops that help the experienced estimator keep up with new technology. Community colleges and technical schools frequently offer courses helpful to estimators.

PERSONNEL POLICIES FOR ESTIMATING

Once a company has developed and trained an experienced staff of estimators, policies should be established to retain these trained men. Good personnel relations involves far more than budgeted expenditures, and providing the proper working tools is an important element.

To remain effective, estimators must keep up with advances in their field. Access to technical and professional journals, reference books, and other job-related literature should be provided. Contact with colleagues from other companies helps estimators see their jobs in proper perspective and learn from one another. Such personnel policies contribute to the type of working environment that promotes maximum departmental efficiency and job loyalty.

COMPUTER APPLICATIONS TO COST ESTIMATING

A number of manufacturers have streamlined their estimating procedures through the use of computers. The extent of usage varies considerably throughout industry, and within large corporations, even among plants.

Without computer assistance, an estimator normally spends only a small portion of his time making decisions and judgments requiring technical skills; data gathering and clerical work such as filing consumes most of his day. Basic estimating tasks calling for experience and skill include:

1) Understanding and/or developing process plans
2) Estimating the time for manufacturing processes and setups
3) Determining tooling costs for alternative manufacturing methods
4) Calculating special costs, such as handling or packaging.

Computer applications to estimating permits the estimator to concentrate upon these facets, thus utilizing his professional capabilities more effectively.

Factors Determining Computer Use

Several key factors determine whether individual estimating departments can make use of computer facilities, and if so, to what extent:

1) Adaptability of product to computerized estimating, which depends upon product size and complexity
2) Speed and storage capacity of available computer
3) Sophistication of available computer programs
4) Availability of computer time
5) Management emphasis upon planning and cost control.

Types of Applications

Computer systems are used by cost estimating in three basic ways: (1) to store data, (2) to perform computations, and (3) to develop complete estimates.

Data Storage. Actual cost data from previous jobs is more accessible when stored in the computer than when filed by conventional means. Under this concept, the estimator obtains processing data, material costs, and labor and burden rates from the computer to apply to the job being estimated. The chief advantage is time saved in searching for data.

Computations. Used as a giant calculating machine, the computer makes short work of mathematical computations which would otherwise consume large portions of the estimator's time. Used in this fashion, the computer saves the estimator's time and assures greater accuracy.

Estimates. With proper programming, the computer system can make the entire estimate. It can determine material requirements, compute machine processing times, and extend these figures by the application of material and labor rates. The program can also provide for tooling costs, general and administrative costs, and a percentage for profit.

Advantages of Computerized Estimating

Well-planned use of computers by estimating departments offer several distinct advantages:

1) Faster estimates. The tremendous speed of modern computer equipment enables estimators to provide the quick estimates frequently needed for contract bids and for management decision making.
2) Greater accuracy. Computerization results in greater accuracy by eliminating the human errors which occur in handling large amounts of data.

3) More economical utilization of plant machinery and equipment. Increased estimating speed allows estimators to consider various alternatives, ultimately resulting in more effective plant utilization.

4) More efficient use of estimating personnel. The elimination of large amounts of routine work permits the estimators to concentrate upon those tasks requiring experience and judgment. The use of computers may also decrease the number of clerical personnel needed by the estimating department.

Computerized Estimating and the Future

Computer manufacturers are exploring and developing various techniques to eliminate much of the paperwork and document handling presently done by estimating departments. With the improvement of computer input/output terminals, and a concomitant decrease in computer costs, estimators may begin storing all estimating data in the computer system. Any portion of this data will be easily retrievable upon demand. Improved input devices now under study include:

1) Optical scanning terminals which read printed data
2) Light pen (to be used with visual display terminals)
3) Audio receivers.

Inexpensive input terminals, readily accessible to the estimator, will replace the adding machines and the row of file cabinets now commonly found in an estimating department. Basic time data used to determine processing times will be stored in the computer memory core, and multiple-use data required by more than one plant or estimating group will be forwarded to a central data bank. Teleprocessing and exchange of estimate information between plants will be more common than at present.

The estimator will have to make the technical decisions before a computer can complete an estimate on a part or process, but the computer program will systematically lead him through the steps necessary to develop it. The image of the estimator and his "little black book" will change as estimators begin spending most of their time making decisions and answering questions under the computer's direction. Through the use of limit tables, the computer will alert him of mistakes in judgment. The results of his original estimate will be readily available at the buyer's desk as required for more effective vendor negotiation.

Manufacturing management will be able to receive their operating reports by part number or operation, the status of each product being automatically adjusted for quantity by use of the learning or progress curve. Without leaving their offices, plant or product managers can request cost projections summarized or in detail according to their needs. As the product design is altered the economic effects of the change can be readily evaluated.

Industrial engineers and load planners can use these projections to evaluate the effects of schedule changes, parts and material shortages, labor force changes, etc. In total, the company or plant management will have almost unlimited capability to evaluate and plan the future, using the basic estimator's input which is dynamically updated as changes in product specifications and design occur. Large computers with massive storage capacity and multiple terminals for input and output will be common throughout industry.

The problems created by this move toward computerized estimating are many. The graduate engineer will no longer be required to estimate costs for the majority of parts or processes. He will either have programming knowledge or will work with professional programmers designing and developing the "look-up tables," decision logic within the computer, and input requirements or computer questions that cannot be structured internally by the computer. The basic estimate will be specifically developed by a technician acquainted with manufacturing processes, but following instructions submitted by the computer and the engineer. Problems will arise in teaching these technicians the coding techniques required by the terminals. However, some opportunity for demonstration of individualism (within controllable limits) must be provided to prevent employees from feeling they are subject to the computer.

Companies will be developing massive standard data banks, formulas, and decision tables within the computer. These data will be frequently tested or altered through simulation studies and continual comparison with actual conditions. Significant areas of deviation will be automatically determined by the computer, and that information will be made available to the estimator and management.

The importance and use of cost estimates will increase with greater computerization. The present problems of the estimator's slow reaction time to the "what would happen if" situations will be almost eliminated; evaluations of this type will be made without the estimator becoming involved.

The estimating engineer will have to be acquainted with the product and with good manufacturing processes and techniques. Because of computers, he will have more time to keep abreast of technological advancements within his field and to assist design engineers to optimize functionality while lowering costs. The estimator will possess an even greater professional knowledge of the processes for which he is responsible, and the monotony and manual manipulation of data will be reduced.

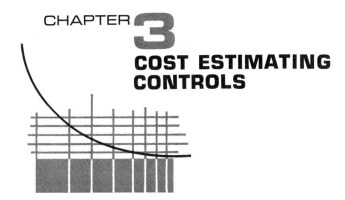

CHAPTER **3**

COST ESTIMATING CONTROLS

Fast, economical, and accurate estimates require proper management control of the estimating function. Management establishes the type of estimating department that will best serve company needs, and then formulates the procedures and administrative controls necessary for efficient departmental operation.

The cost estimating department constantly interacts with other departments, necessitating controls over paperwork and other forms of communications. For example, cost estimate requests can be initiated by the marketing, product engineering, manufacturing engineering, and cost-reduction functions. The purchasing, process planning, methods engineering, plant engineering, and accounting departments then supply the estimating department with processing and cost data. All of these interactions require communication and control.

The estimator himself establishes certain controls to increase the accuracy of his estimates. This can be accomplished by estimate simplification and other procedures discussed later in this chapter.

ADMINISTRATIVE CONTROLS

Desirable administrative control measures include monitoring of incoming cost requests and the establishment of routings for cost requests and completed estimates.

Monitoring Cost Requests

Regardless of the size of a company, all cost requests should be examined before being forwarded to the estimating department. A department or person should be designated to check each cost request to see that the product for which an estimate is desired is consistent with the company's product mix, production schedule, equipment capacity, available labor skills, and company policy. Such

19

a screening procedure reduces the cost estimating department's workload by allowing estimators to develop cost figures for only those jobs that the company is capable of, or desires to produce.

An individual or group responsible for soliciting business for the company, usually the sales department, is the logical choice to carry out this screening. The estimator does not, in the interest of making better use of his time, ordinarily question the validity of a cost request coming to him. However, he is often consulted as to feasibility by departments requesting cost estimates before the department releases the request.

Many companies bid on all requests for quotes out of courtesy. Though difficult to segregate, the cost of estimating such requests is actually a customer relations cost.

No company is capable of making every product, but even though it seems uneconomical to quote on a new item that can be easily underbid by a competitor experienced in that field, there may be some merit in the interests of expanding a product line and in quoting a potential product to determine whether the new or special equipment and skills needed to produce it should be acquired. This is a matter of policy with which a request for cost estimate should be consistent before it is acted upon. An unfamiliar product area should be investigated very carefully before a cost quotation is released because one unforeseen cost can quickly reduce or eliminate the profit margin.

Routing for Cost Estimates

As shown in Fig. 3–1, the general path of a cost request through a typical company to arrive at an estimate is as follows:

1) A customer submits an order or bid request to the marketing department.
2) Marketing or other authorized personnel screen the request and, if approved, sends a request for an estimate to cost estimating along with all available data, e.g., drawings, specifications, and volume requirements.
3) Cost estimating gathers all pertinent costing information and writes up the cost estimate on suitable forms, including, perhaps, tentative operation sheets; material, labor, and tooling calculations and summary sheets for each; and master summary sheets. In collecting processing and cost data, the estimator works closely with five other departments: (a) purchasing, (b) process planning, (c) methods engineering, (d) plant engineering, and (e) accounting.
4) Cost estimating forwards the completed estimate to the proper authority for checking, adjustments, comments, and approval.
5) The approved estimate is returned to the marketing department with the initial request in one of two ways. The person responsible for final approval may route the cost figures directly to the marketing department or he may return the estimate to the estimating department for transmittal to marketing.

   In either case, the estimating department should receive a copy of the approved estimate because any changes or additions reflecting policy will affect future estimates. The estimating department files all worksheets, summary sheets, etc., for future reference; a well-planned filing system is extremely useful.

6) The marketing department informs the customer whether the company can furnish the product at the customer's desired price or informs him of the actual estimated price, tool cost, and lead time applicable for the desired quantity.

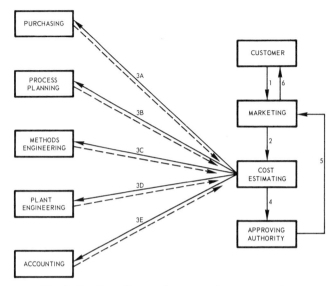

Fig. 3-1. Flow diagram for cost estimate processing.

If a request is originated by a department within the company, Steps 1, 2, and 6 are eliminated, and the originating department sends the request with all available descriptive data directly to cost estimating.

When departments other than cost estimating participate actively in a cost estimate, Step 3 breaks down into 5 sub-steps: (a) the list of purchased parts and materials is routed to purchasing for a cost, (b) the drawings and specifications go to process planning for preparation of a manufacturing plan, (c) methods engineering estimates manufacturing time, (d) facilities and material handling costs are determined by plant engineering, and (e) labor costs are entered by the accounting department. The cost estimator collects and summarizes these costs, and after the estimate is properly reviewed, it is forwarded to the responsible authority for approval, or returned to either the marketing department or directly to the originating department.

INITIATING COST REQUESTS

As discussed in Chapter 2, the estimator may report to any one of several department heads. Only rarely, however, does the major portion of his workload emanate from the department to which he reports. Although any manufacturing department may initiate cost requests, most (except in the process industries) originate in the marketing department.

Next in volume of cost estimate requests are manufacturing engineering, product engineering, and such departments as cost control, value analysis, and value engineering. Marketing, manufacturing engineering, and product engineering are in close contact with customers or potential customers, and rely upon the estimating department for realistic cost figures upon which to make decisions as to whether to accept customer orders, make changes in manufacturing processes for existing products, or develop new products. Cost-reduction functions rely upon the estimating department to validate cost reduction proposals.

Cost Requests from Marketing

Marketing may request cost estimates for one of two reasons: to fulfill a specific customer's requirements, or to investigate the manufacturing feasibility of a new product.

In the first instance, the marketing department receives a request for a quotation from a potential customer, screens the request to assure that the company is interested in presenting a quotation, and requests a manufacturing cost estimate from the estimating department. In the second instance, marketing may also request a cost estimate for a new product invented or developed for sale to potential markets. In this case, the marketing department would use the resulting estimate to ascertain whether the product could be manufactured and marketed profitably. In either case, the marketing department may request estimated costs for varying production levels, or may want to know what lot size can be manufactured at a predetermined cost.

Fig. 3–2 shows a sample form used by one marketing department to request cost estimates. The sales representative fills in this form and attaches any drawings, specifications, or other pertinent data he receives from the customer. Any prototypes, dummies, or other physical models supplied by the customer would also accompany the form.

REQUEST FOR ESTIMATED COST

SHEET ______ ESTIMATE NO. ______

TO: DATE: ______

FROM: ______

______

AN ESTIMATED COST TO SUPPLY THE GOODS AND OR SERVICES BY YARD-MAN, INC., TO THE CUSTOMER AS DESCRIBED BELOW IS HEREWITH REQUESTED.

ITEM NO.	MODEL OR PART		ENG. CHG LETTER	QUANTITY REQ.		JOB START DATE	COMMENTS
	NUMBER	NAME		MONTH-20 DAY.	ANNUAL TOTAL		

NAME OF CUSTOMER ______ ADDRESS ______ TEL. NO. ______

NAME OF CUSTOMER CONTACT MAN ______ ADDRESS ______ TEL. NO. ______

YARD-MAN, INC. TO FURNISH EVERYTHING EXCEPT: ______

______

YARD-MAN, INC. CONTACT MAN ______ ADDRESS ______ TEL. NO. ______

COMPLETED ESTIMATED COST DUE DATE: ______

APPROVED: ______
PRESIDENT

Fig. 3–2. Cost estimate request form used by the marketing department.

Cost Requests from Product Engineering

This group uses cost estimates as an aid in choosing the most economical design for a given product's function or quality by comparing estimated manufacturing and material costs for alternative designs. Product engineering also initiates cost estimates to remedy field installation difficulties or to correct product deficiencies. Estimators compile such estimates from detailed layouts, sketches, or samples of the assembled product or parts. Layouts need not be completely detailed, but the greater the detail, the more accurate the estimates. Material, tolerances, and surface finish specifications must be established within reasonable limits because these factors greatly influence costs.

Fig. 3–3 illustrates a typical request for a design cost estimate.

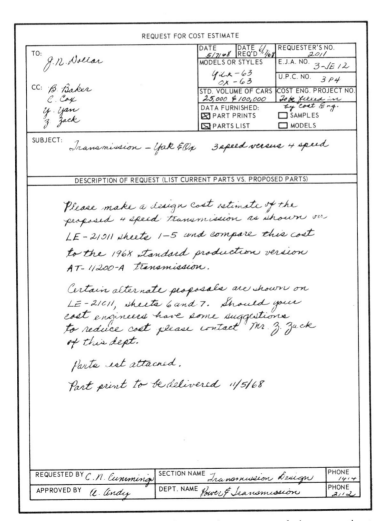

Fig. 3–3. Cost estimate request form used to request a design cost estimate.

Cost Requests from Manufacturing Engineering

Process planners from the manufacturing engineering department use cost estimates to establish the manufacturing costs for alternate manufacturing plans and for selecting the most economical manufacturing method and tools based on a standard volume. The data may also be used to approximate the break-even points of the alternate plans for varying production quantities and rates.

Costs Requests from Other Departments

The estimating department also receives estimate requests from cost reduction groups and from purchasing to assist in evaluating vendor quotations.

Cost Reduction. Cost control, value analysis, and value engineering departments also initiate cost estimate requests to help them scrutinize production parts and discover ways to reduce costs without lowering quality or functionality. Almost all suggested changes in product design or manufacturing techniques

COST REDUCTION PROPOSAL

TO: Mr. J. N. Dollar
CC: Mr. B. Baker

DATE 6/17/68
ORIGINATORS FILE NO. 12010
MODELS AFFECTED All 1968
CT NO. A-012
UPG NO. AZA

DATA FURNISHED:
[X] PART PRINTS
[X] SKETCHES
[ ] LAYOUT
[ ] SAMPLES
[ ] MODELS

SUBJECT: Impeller - current production cast iron. IT-1606-A versus proposed die cast design.

DESCRIPTION OF PROPOSAL

Please prepare a manufacturing cost estimate of the proposed die cast impeller versus the production cast iron impeller IT-1606-A. Tool and facilities cost should be included in this analysis.

REQUESTED BY C. Kost	SECTION NAME Production Cost Red.	PHONE 444
APPROVED BY R. D. Pruit	DEPT NAME Administration	PHONE 445

LEAVE BLANK TO BE COMPLETED BY COMPETITIVE PRODUCT ANALYSIS DEPT

COST DATA

.65 COST PER CAR INCREASE [ ] DECREASE [ ] $.15 TOOL COST $20,000 FACILITIES None

CONTROL VOLUME 150,000 MONTHS TO AMORTIZE TOOL COST 7½

PER K Kutz DATE 11-14-68

Fig. 3–4. Form used by a cost-reduction group to request a cost analysis.

will affect cost so that cost estimating plays a vital role in establishing the soundness of a proposed change. Fig. 3–4 shows a sample form used by a cost-reduction group to request a cost analysis, as well as submit such a proposal to management.

Evaluating Vendor Quotations. The purchasing department requests cost estimates on material or parts it intends to buy. For parts, a detailed estimate covering materials, tooling, and labor is needed to determine whether quoted prices are realistic. Savings may also be effected by comparing quotes on material and component costs. This allows purchasing to negotiate with vendors for more favorable prices if a particular quote is too high and provides a double check on all vendors. A comparison of estimates between the vendor and the purchasing company may suggest cost-saving measures as well. Tooling costs, for example, may be reduced by lending tooling to suppliers.

Fig. 3–5 shows a form that can be used to analyze vendor quotations and re-

VENDOR COST ANALYSIS

TO: Sales Dept.
Jim Hill
FROM: Cost Est.
L. Backus

DATE	DATE REQ'D	REQUESTOR'S NO.
2/29/68	3/15/68	X-14376-Y

MODEL OR STYLE	VOLUME
COB-X-3-RA	30,000/mo.

DATA FURNISHED:
☒ PART PRINTS ☐ SAMPLES
☒ PARTS LIST ☐ MODELS

SUBJECT: Front bumper and guards

	VENDOR COST	COMPANY COST	DIFFERENTIAL	COMMENTS
MATERIALS	2.713	1.978	–.735	check to see if they are getting quantity disc.
LABOR	1.216	1.450	.234	Non-union shop
TOOLING	2.760	1.100	–1.660	They're figuring on new tools
TOTALS	6.689	4.528	–2.161	

SUGGESTED ACTION: Their labor is better. We can probably work out something to cut tooling costs - (We have a setup in the shop already.) They're way off on material - may not be figuring in the volume disc. Suggest buying part if they can get price down to 4.5 or lower.

Fig. 3–5. Form used by estimating department to analyze vendor quotations.

port results to requesting departments so that a make-or-buy decision can be made.

ESTIMATING METHODS

Estimating may be accomplished by: (1) conference, (2) comparison, or (3) detailed analysis. While the first two methods are satisfactory when only a preliminary or rough estimate is needed, final cost estimates should be made only through detailed analysis. Only detailed analysis assures the accuracy needed to estimate a major manufactured product upon which thousands of dollars may depend.

Conference Method

In this method, representatives of various departments such as purchasing, process planning, tool design, and time-and-motion study confer and estimate the costs of material, labor, and tooling. A coordinator from the estimating department collects these costs and applies burden factors to develop a total manufacturing cost for the product.

The conference method may also be used within the estimating department. Estimators having specialized knowledge confer on an estimate and determine a cost figure without needing counsel from other departments.

The chief advantage of the conference method, and the main justification for its use, is its speed. It permits experts in different fields to pool their knowledge when quick estimates are needed. The method is also useful to companies not having established estimating departments, and who still must develop product cost data.

The conference method's main disadvantage is its lack of accuracy; resulting cost data should be treated cautiously and checked meticulously. The accuracy of any estimate depends upon the availability of specifications, drawings, and samples, and in actual practice these are seldom obtainable on short notice when quick estimates are desired.

Comparison Method

The comparison method relies on an accumulation of past experience and data. The estimator applies up-to-date costs derived from similar parts to the project, adjusting these costs to suit material, labor, and processing variations. Caution should be exercised in using data compiled from products manufactured in larger or smaller quantities and the estimate should be factored accordingly. While some of the elements will remain fairly constant, others will change. Material usage, for example, may remain fairly constant, though slightly lower spoilage rates sometimes result with increased production runs. Larger production lots also decrease setup costs and offer the opportunity for more efficient use of durable tooling. Labor costs similarly decline with increased efficiency developed over longer production runs.

Another way of using the comparison method is the application of a rate per unit of measure factor. The unit rate, based upon actual production data, may be hours per pound of material, dollars per cubic foot, etc. Careful judgment must be used in applying such rates because of the cost changes to which they are subject.

Like the conference method, the comparison method is used when time is short, and the two methods are often used jointly.

Detailed Analysis Method

Detailed analysis, which is generally performed by an estimator working alone, is the most reliable method of estimating. As its name implies, detailed analysis includes a complete examination of all the important factors involved in the production of a manufactured item. Historical data are used, but only after validation. Because estimating by detailed analysis requires strict adherence to the procedure described below, it is more time-consuming than the other two methods. Fig. 3–6 shows an example of an estimate prepared by the detailed analysis method.

PART NO	PART NAME	STD HRS	MAT'L	LABOR $2.60	VAR BURDEN 200%	TOTAL VAR COST	FIXED BURDEN 20%	TOTAL COST	TOOLING
M-40040	Motor Mount								
Material	.079 A.I.S.I. C-1008 Cold rolled by 1000' coil by 36 x 1/2 wide								
	Length Multiple = 5" Width Multiple = 9"								
	WT/Coil = 11,200 Pcs/Coil = 9600 wr/pcs.=1.6600@ $.0725/lb.		.0845					.0845	
	Total Operation	.01290		.0335	.0670	.1005	.0067	.1072	
	Cutoff and draw	.00200							$6,000 00
	Flange	.00200							3,850 00
	Trim and Pierce	.00200							5,450 00
	Wash, paint and Pack	.00500	.0155					.0155	
	Setup	.00100		Tools = 15,550 Volume =100,000 Unit Tool Cost = .1555 Value of Scrap = .0008 Equip = Available Rearrangement: None				Gauges	250 00
	Inspect	.00090							
	Totals		$.1000	$.0335	$.0670		$.0067	$.2072	$15,550 00

Fig. 3–6. Detailed analysis cost estimate for a motor mount.

Each of the following steps must be performed when preparing a detailed analysis estimate:

1) Calculate raw material usage, including scrap allowances and salvage material (direct material)
2) Process each individual component (write the operation sheet)
3) Compute the production time (direct labor) for each operation
4) Determine the equipment required (new, rework, or on hand)
5) Determine the required tools, gages, and special fixtures
6) Determine any additional equipment needed for inspection and testing.

While the conference or comparison methods are satisfactory for preliminary estimates, detailed analysis is used by almost all manufacturers despite the extra work and additional time required for its completion. This method furnishes the most accurate prediction of anticipated costs on new products or for engineering changes on currently manufactured parts.

CONTROLLING THE COST ESTIMATE

The purpose of the estimating function is not to produce exact cost data but rather to supply cost figures having a high probability of falling within an acceptable range. Therefore, product cost estimates seldom coincide with actual manufacturing cost figures, and neither the preliminary estimate nor the final estimate is expected to be exact. Preparation of precise cost figures would be excessively time consuming and expensive, even if they were possible to predict.

However, greater accuracy is required for some estimates than for others. Parts or products expected to be produced in large quantities over long periods of time should be estimated as closely as possible because their potential profit or loss is much higher than items of the same cost manufactured in small quantities. Similarly, expensive items, even though produced in small lots, deserve careful costing.

Deviations between estimated and actual cost figures result from several factors. Human error is a major cause but, as was pointed out in Chapter 1, in some cases estimators purposely do not analyze in detail every item affecting cost because to do so would waste valuable time. Unpredictable variables in the manufacturing processes and changes in material costs can also cause significant deviations.

Estimators should try to be aware of each of these factors when costing products and attempt to control them. Direct control measures include the use of machine and worker performance factors, estimate simplification, and indexes which reflect cost changes.

Deviations in Cost Estimates

An estimated cost deviates from the actual cost of an item because all the factors affecting the cost cannot be fully evaluated. The largest and most important factors are given primary attention, but many factors of lesser importance must be left to influence the results in a random manner. Some factors cannot be predicted with certainty at all such as delays caused by defective material and machine breakdowns. An equipment performance factor will spread such losses over all jobs with the result that, when equipment and machinery runs smoothly, estimates will be high, but when breakdowns occur, estimates will be low.

Fig. 3-7 shows how estimated costs can deviate from actual costs. The error curve resulted when the per cent error of each of 157 estimates of the labor cost of making tools was calculated. Each bar represents the number of estimates within a given range of 10 percentage points of error. The deviations range from 50 per cent low to 450 per cent high. If an infinite number of estimates were plotted and the pattern remained approximately the same, the distribution of errors would be represented by a curve such as the one superimposed on the histogram in Fig. 3-8.

The estimates upon which Fig. 3-7 is based were made when an abundance of orders was available and the estimator was able to estimate doubtful jobs safely. As a result, the curve is skewed to the high side. However, most of the

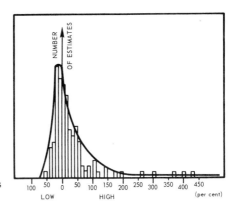

Fig. 3-7. Deviation of estimated costs from actual costs.

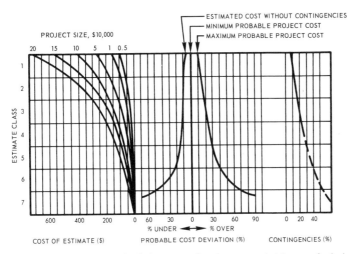

Fig. 3-8. Hypothetical curves depicting cost of estimate, probable cost deviation, and a contingency factor.

estimates show relatively small deviations, which is a general characteristic of good estimating practice.

Average Estimates Versus Actual Costs

Although cost estimates are not expected to conform to actual manufacturing costs, the average of the estimates over a period of time must be reasonably close to the average manufacturing cost. Probable cost deviations, correlated with estimated costs and contingency factors, are depicted as curves in Fig. 3-8. Such a comparison can only be made when most of the cost-estimated jobs or products are manufactured in-house, thus establishing a verifiable actual cost. Even if identical information is available for two jobs (either similar or dissimilar) and the same method is used to estimate both jobs, the estimates can differ from the actual costs of the jobs. Unforeseen problems may arise during one job, while during another job no difficulties are encountered.

The difficulties may include faulty tooling, machine breakdown, material short-ages requiring substitution of a more expensive material, or labor problems.

Low Estimates. The most significant causes of low final cost estimates are:

1) Higher labor and material costs than anticipated
2) Incorrect design information
3) Unexpected delays resulting in premiums paid for overtime and materials
4) Unexpected processing problems requiring deviation from the pre-liminary manufacturing plan
5) Failure to rework the preliminary estimate to produce an accurate final estimate
6) Inability to meet part print specifications using the material specified.

Every underestimated job represents a potential loss to the company. Al-though these losses may be balanced by gains from high estimates, this results in a break-even operation, not a profit-making one.

High Estimates. High estimates may be caused by:

1) Overestimating bid requests that the company does not want
2) The tendency of estimators to be overly cautious when a job requires processes that they are not familiar with
3) Making a "guesstimate" and then increasing it to cover contingencies
4) Planning for more processing steps or a higher level of tooling than is actually required
5) Failure to take advantage of "price breaks" on quantity purchases of material
6) Overestimating labor costs by failing to take into consideration tasks that can be performed by operators while machines are in cycle.

Even though a high estimate may result in greater profit than anticipated, these estimates are also undesirable because they:

1) Lead to overbidding and cause rejection of proposals that could be profitable
2) Cause loss of customer good will if quotations are consistently high
3) Lead to waste in design and fabrication when they become the initial standards of performance
4) Require a larger staff of estimators because a greater number of esti-mates must be performed for a given volume of completed jobs.

Controlling Estimate Deviations

Cost estimates are characterized by two types of errors: random and biased.

Random Errors. Some cost estimating errors occur at random and their frequency can be high or low. Causes of error are comparable to assignable and chance causes in statisitical quality-control work. In quality control, the methods of correction are to find and eliminate, or minimize, the assignable causes, and also measure the probable effect of, and control, the chance causes. In this manner, the true quality level and the likely quality deviation in a manufac-turing operation can be determined and a warning given immediately when conditions are out of control, making losses imminent. The same methods can be utilized by cost estimating where sufficient historical data have been accumu-lated.

Biased Errors. Some errors in cost estimating are biased, i.e., they follow trends which may be due to assignable causes. Three common examples of such trends are:

1) Fluctuation in labor and material costs with economic conditions
2) The cost of a machine, tool, or piece of equipment usually varies with size or capacity; larger sizes can be expected to cost more than smaller ones but not necessarily in direct proportion
3) The amount of time, and therefore the cost of performing an operation (particularly one requiring much manual time), decreases as the number of units produced increases.

A good estimator should find that the sum of a large number of his estimates is close to the sum of the actual costs of the jobs performed. Most estimators check their performance over various periods of time to determine the percentage by which their cost estimates deviate from actual manufacturing costs. Once they are able to predict their estimating bias, they modify future estimates by means of a personal performance factor.

Project Simplification. Estimating errors are normally smaller when the elements of a project are estimated individually than when the project is estimated on an overall basis. Likewise, the errors made in estimating for small and simple elements are fewer than those for large and complex elements.

Use of Equal Size Elements. A project must be split into elements of approximately equal size to benefit fully from estimating simplification. If a major part of a product is not closely estimated and minor details are thoroughly estimated, the sum of the two will often result in an inaccurate total estimate. Any refinement gained by thorough estimating of the minor parts is futile because it cannot offset the effect of a cost error made when estimating the major portion.

For example, the cost of converting a regular machine tool headstock to a special headstock is to be estimated. Because the conversion requires only the replacement of a few parts in the standard headstock, a reliable estimate can be obtained from the established price of the standard headstock and a detailed study of the changes required. On the other hand, if an entirely new headstock is to be estimated, it must be broken down into elements of approximately equal size to avoid sizable cost errors.

Good estimating practice recognizes the need to build up an estimate from approximately equivalent elements, as indicated by the chart in Fig. 3-9. Common practice in tool estimating is to determine the operation time and material for each part (without considering possible material losses) and the overhead from one or more rates. In estimating the products manufactured in quantities, the manufacturing operations are divided into elements and all factors affecting material and labor costs are considered, and overhead is allocated according to the kinds of operations required to produce the item.

Estimating in a Changing Cost Environment

Cost variations will obviously affect the estimate. If an item is to be manufactured within a few days or weeks, current costs for labor, material, and overhead are usually safe. On the other hand, using such current costs for estimating may be hazardous if actual costs are to be incurred and returns realized

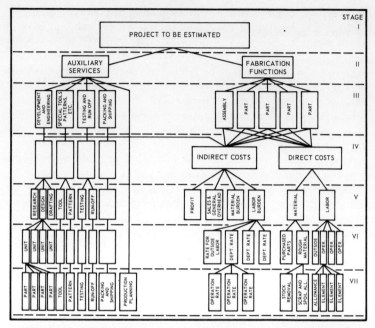

Fig. 3-9. The progressive stages into which a project can be divided.

after a longer time lapse. If costs should increase over a period of time due to additional planning and designing for example, a serious estimating error and consequent financial loss may result. In contrast, should prices decline, a product may not be marketable at the previously estimated profit.

Likewise, a proposed project may be similar to one already completed for which cost figures are available. During the intervening period, however, the basic labor, material, and tooling cost rates may have changed appreciably. The extent of such changes must be known in order to apply the previous project costs to the current project. Updating past costs to present cost levels requires techniques similar to those used in predicting future costs as described below.

Estimates for Future Production. Present costs and quotations must serve as starting points in preparing estimates for the future, but they may need to be modified by forecasting conditions at the expected time of manufacture. An estimator must be aware of economic trends or seek advice from a person qualified to predict price trends and market stability.

In preparing estimates for future production, not only future prices, but anticipated volume and facilities must be studied. At a lower production volume than expected, fixed and overhead costs per unit increase. A higher output may not only alleviate fixed charges, but also permit utilization of more efficient work methods. Additional lines of products may absorb plant overhead costs, or vice versa. As noted earlier, estimates for future production are generally made only for guidance or planning purposes, and are followed up by detailed estimates.

The estimator cannot recognize cost trends without an understanding of the principles behind the cost accounting system furnishing him with information. Having such an understanding, an estimator is able to appreciate why overhead rates are changing, the relationship of overhead to shop activity, labor costs, and the general production picture.

Cost Indexes. In certain circumstances, one or more of the many published price or cost indexes may be helpful in evaluating past changes in costs or forecasting future trends. A price or cost index consists of a series of index numbers, each calculated to show the price level for a particular time period in relation to the level at a reference or base period.

As an example, assume that a tool shop successfully coordinated its cost figures with one of the published indexes. If a change of one point in that index coincides with a change of 75 per cent in tool costs, the cost of a similar tool can be easily adjusted to current prices. For instance, assume that a certain tool cost $1,000 in July, 1966, at which time the index was 169.3. The tool cost is to be adjusted to a figure for December, 1968, when the index is 180.6. The adjusted tool cost can be determined as follows:

$$1,000\left[1 + \left(\frac{180.6 - 169.3}{100} \times .75\right)\right] = \$1,085$$

In using a cost index for estimating, the estimator should keep a record of the index for the month or other period of time that he has used in his estimate. Periodically, the actual costs must be compared with the estimated cost to evaluate the accuracy of the estimate. Also, changes in actual costs must be compared with changes in the index so that adjustments in the factor can be made.

If published indexes do not fit the specific needs of a given company, the estimator must prepare his own index based on his historical data file. Once the index has been compiled, adjustments in cost can be made to conform to it as changes warrant. Estimators concentrating in specific manufacturing areas will soon recognize the cost increase percentages of tooling and other items affecting their products and can factor these data into their indexes for ready reference.

DO'S AND DON'T'S OF COST ESTIMATING

The following points should be kept in mind when making a cost estimate:

1) Do use a printed form wherever possible to assure completeness in estimating. Continuously strive to devise better printed forms, or new forms for similar estimating tasks that recur to eliminate the necessity for developing special forms for particular estimating situations as much as possible.

2) Do reduce computations to a minimum by assembling standard data in well-organized files. Supplement departmental files by adding new standard data at every opportunity.

3) Do have your computations checked by your supervisor before release. It is difficult for an estimator to check his own work, and if not corrected the first time, the same error can perpetuate itself indefinitely.

4) Do review required tolerances on engineering specifications carefully. If tolerances are not given, query the originating department.

5) Do verify the extent of similarity between two jobs before reusing previously estimated or actual job data for a second estimating assignment. Be especially careful of material, tolerance, and quantity requirements.

6) Do keep searching for new materials, processes, machines, and techniques to improve present manufacturing procedures.

7) Do keep a properly identifiable set of notes for each estimate; include the blueprints, specifications, correspondence, and other material related to the job.

8) Do remember that much of the data and information contained in an estimate is confidential and that business is highly competitive. Try not to help a competitor by carelessly revealing the valuable information contained in an estimate.

9) Do get as many quotations for purchased parts as time permits.

10) Do make comparisons between the cost of buying a particular service and modifying company facilities to perform the operation.

11) Do not depend on "hunches" to yield a reliable estimate; analyze each step carefully.

12) Do not guess at the meaning of unfamiliar specifications or symbols. Check with the customer or a reliable source for an interpretation and get the definition in writing.

13) Do not base an estimate on the projected purchase of new equipment unless the equipment will pay for itself within a reasonable period of time.

14) Do not apply high-production manufacturing methods to low-production jobs or vice versa.

15) Do not assume that specifications are final. Unless the customer specifies no product design changes, ask for any changes that will reduce manufacturing costs and save money for the customer.

16) Do not accept verbal engineering changes or similar directives; insist on written documents and file them in the appropriate folder with all other relevant data.

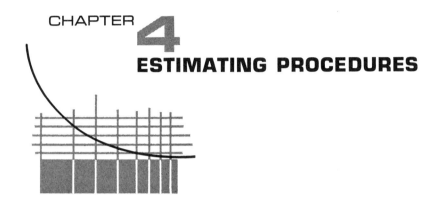

CHAPTER 4
ESTIMATING PROCEDURES

The best estimating procedures for any estimating department are those that can most satisfactorily meet the manufacturing organization's needs for cost data. These procedures vary with the manufacturing organization's size and complexity, with the sophistication of the firm's products, and with the variety of its product mix.

Estimators in small plants manufacturing relatively simple products usually rely on uncomplicated techniques. The owner-operator of a small job shop, if sufficiently experienced, may scan a part print and scribble cost per unit figures directly on the drawing. The small size of his organization permits simple estimating procedures, and his uncomplicated products and product mix allow his memory to serve as an adequate filing system.

At the other end of the spectrum is the manufacturer with thousands of employees producing a wide variety of items in long and short production runs. Estimators in operations of this type must follow prescribed procedures designed to cope with many complicated situations in order to provide accurate cost data for customers, as well as in-house departments requiring various kinds of anticipated cost data for decision-making. Efficient operation requires formalization of estimating procedures and maintenance of extensive records to ensure continuous profitable operation even though personnel and products change.

Despite the gap in applicability and philosophy between the two modes of estimating operations described above, profit is the final goal of each. The size of the estimating departments may vary, but estimators in any size or type of operation must perform certain basic steps to establish accurate cost estimates:

1) Analyze the cost request to ensure that all essential information is included and to determine what is actually being requested
2) Analyze the part or product and list all standard parts and the parts to be fabricated
3) Understand the manufacturing process plan established for each part

35

4) Compute the material costs for the standard and fabricated parts
5) Determine the costs of durable and perishable tools
6) Estimate the manufacturing time for each operation listed in the manufacturing process plan
7) Apply the labor and burden rates to each operation. In some companies this step is performed by the accounting department.

It is difficult to generalize on the amount of time necessary to produce a good estimate. Although estimating accuracy and speed are largely a function of the estimator's skill, there is no doubt that an estimate improves as more time is allowed for completion. However, Parkinson's law—that work expands to fill the time allocated for it—applies to estimating as to any other function (1). A good balance must be struck between the requirements of accurate estimates and the necessity for quick completion to meet the need for cost data.

If the estimator does not have cost data on file for material, tooling, equipment, standard parts, and purchased parts, he must use valuable time to obtain these figures from the responsible departments. As the time required for an estimate varies, so does the time required to accumulate the pertinent figures from supporting departments. The experienced estimator will take preventive measures, such as building up a file of historical cost figures to reduce his reliance upon vendors and other departments and to speed up his estimating procedure.

ESTIMATE ANALYSIS

Upon receiving an estimating assignment, the estimator should first analyze the cost request and carefully inspect all accompanying material. This examination acquaints him with the part to be estimated and alerts him to any missing information. He should check the request for each of the following items:

1) Estimate due date—on overdue cost estimate requests, check back with the requesting department to determine another date, if any
2) General design specifications—a brief description of the product, its function, performance, and purpose, e.g., a pump to lift a highly corrosive fluid 20 ft. at a rate of 10 gpm.
3) Quantity and rate of production
4) Assembly or layout drawings which will usually include only overall or general dimensions. Processes, raw materials for the parts, dimensions, and tolerances are not usually specified.
5) List of the proposed subassemblies of the product
6) Detail drawings and a bill of materials for the product
7) Finishing requirements for the part or product
8) Test and inspection procedures and equipment
9) Detailed machine tool, tool, and equipment requirements
10) Packaging and/or transportation requirements
11) Manufacturing routings
12) Standard time data
13) Material release data
14) Subcontractor cost and delivery data
15) Area and/or building requirements.

The greater the number of the above items and the more complete each item is made available to the estimator, the more accurate the estimate can be. Usually, however, the more accurate the cost estimate, the greater the expense involved in making it. As shown in Table IV-1, for a Class 1 estimate, *all* pertinent information is taken into consideration when making the estimate. At

Table IV-1. Cost Estimating Data.

Description of Data	Class of Estimate						
	1	2	3	4	5	6	7
General design specification, quantity, and production rate	x	x	x	x	x	x	x
Assembly or layout drawings	x	x	x	x	x	x	
Proposed subassemblies	x	x	x	x	x		
Detail drawings and bill of material	x	x	x	x			
Test and inspection procedures and equipment	x	x	x	x			
Machine tool and equipment requirements	x	x	x	x			
Packaging and/or transportation requirements	x	x	x	x			
Manufacturing routings	x	x	x				
Detailed tool, machine tool, gage, and equipment lists	x	x					
Operation analysis and workplace studies	x	x					
Standard time data	x	x					
Material release data	x	x					
Subcontractor cost and delivery data	x	x					
Area and building requirements	x						

the other extreme is a Class 7 estimate in which *only the general design specifications and quantities* are considered. A Class 7 estimate must be based on the estimator's knowledge of the product or on a comparison with a similar product.

The classes of estimates differ, not only in the amount of information needed, but also in the method by which the estimating is done and the time available to the estimator. Class 1 and 2 estimates, for example, depend upon firm quotations of machine tools, equipment, tools, and materials, and a detailed synthesis of all operations. For the Class 2 estimate, note that all information is available for estimating the product except the area and building requirements.

PART ANALYSIS

Departments requesting cost estimates should include a complete bill of materials whether the engineering specifications for the part or product originate in-house or elsewhere. When a list of materials is not provided by the initiating department, the estimator must compile one from whatever drawings he is given. A bill of materials is little more than a detailed listing of the components of the manufactured product, but it serves several purposes:

1) As a checklist to prevent omission of details
2) Permits a segregation of standard purchased parts from those to be fabricated
3) Lists all of the parts to be estimated.

Make-or-buy decisions are made from this detailed listing. Small standard items such as bolts or screws are generally purchased, as are parts too large, too small, or otherwise unsuitable for existing plant equipment. Parts requiring special processing, such as heat treating, are also often subcontracted.

Subcontracting

Even though the proper equipment may exist in-house for a particular job, it is sometimes advisable to subcontract it for one of two reasons:

1) The shop load may be such that a component cannot be made internally in time to meet a customer's requirements.
2) The profit margin for a particular component, when subcontracted, may be greater than if manufactured internally, particularly when sufficient other work is anticipated to keep the equipment running.

The latter reason for subcontracting usually requires a decision from an authority other than the estimating group, and this requires time. Where time is adequate, however, the responsible authorities should be made cognizant of the subcontracting advantages. When time does not permit the referral of such decisions, the estimator must assume that the component will be made in-house.

Comparison Method

In reviewing the detailed bill of materials, the estimator should check previous estimates for similar parts in order to expedite the procedure. In making comparisons, a good filing system for past estimates is very helpful.

Suggesting Changes

How far should an estimator go in suggesting changes in specifications in order to reduce cost and facilitate manufacture? Certainly, as one of the reviewers of the engineering drawing, he should note obvious errors and make sound suggestions to correct them. Ordinarily, however, the designer has already reviewed all points requiring detailed analysis and has worked out the best possible solution.

How far the estimator should go to verify or corroborate his suggested changes depends on the amount of time he has to complete the estimate and the amount of money involved. When his suggested change is significant, it should be submitted to the designer for review. Although the estimator should be capable of observing cost reduction potential from an engineering drawing, his time (at least theoretically) is more constructively contributed by actual estimating, leaving value analysis, utilizing his suggestions, to others.

PRELIMINARY MANUFACTURING PLAN

Determining the cost of the processing steps involved in fabricating a part or assembly to conform to specifications is one of the most demanding estimating tasks. Each part must have a processing plan which is usually drawn up by process planners within the manufacturing engineering department. The estimator must be familiar with the tooling, equipment, and time necessary to produce the part to the blueprint. Only the operations requiring direct labor should be listed, and the machines should be identified only by group unless a special machine within a group must be used. This operation lineup is then reviewed by the estimator to determine tool costs.

In a small shop, the processing plan may be developed by the estimator and may involve only a mental review. In order to do this, the estimator needs to maintain good liaison with manufacturing engineering to assure that he is estimating on the basis of the latest and best manufacturing techniques. In small plants that do not have a manufacturing engineering department, the estimator, himself, will have to keep up with current manufacturing methods.

However, as the plant increases in size and complexity, the preliminary process and tool concept should be developed and recorded in orderly fashion for the estimator by the manufacturing engineering group.

If a plant is operating as a job shop and therefore handling a substantial volume of estimates, it is uneconomical to work out the processing plan in the detail that is necessary for a manufacturing release because, in some companies, as little as 10 per cent of the work estimated will return as orders. This indicates that the preliminary process plan should be in a brief form, yet sufficient to maintain estimating accuracy. If there is a low volume of estimating that does not occupy one man's full time, a process engineer may do the estimating.

FACILITIES ESTIMATING

With the completion of a preliminary process plan, it may be necessary to obtain a facilities cost estimate because new products sometimes dictate re-arrangement of machines; relocation of outlets for such utilities as gas, steam, water, and electricity; and even alterations to the building structure. New processing equipment such as paint spray booths, furnaces, ovens, hoppers, and overhead conveyors may have to be designed and built. All of these items are classified as facilities costs.

In some companies the product cost estimator develops the costs associated with additions and changes to plant facilities. In other companies the estimator relies upon the facilities planning group for this cost data. In the latter case, the estimator makes a formal request for a facilities estimate, presenting a written plan showing what equipment has to be moved and indicating any other necessary alterations. New equipment to be purchased is listed.

The same steps must be followed regardless of who does the estimating. The individual must know what equipment is to be installed and what its requirements are in terms of space, utility services, and materials handling equipment. Then he must analyze: (1) the areas where the equipment is to be placed, (2) the existing facilities, (3) the required facilities, and (4) the new product flow pattern. From this analysis, a preliminary plant layout can be made and used as a basis for the estimate. The completeness of the layout is determined by the size and complexity of the facility changes, the required accuracy of the estimate, and the estimate lead time.

After the layout is complete, each piece of equipment is analyzed to determine what is required for its installation. This step is similar to preparing manu-facturing processing or operations sheets for a part or product, i.e., the estimator determines what operations must be performed to obtain the desired results.

To make a reliable estimate of new equipment installation, the estimator must also add the labor rates for all trades and skills involved and the cost of materials to be used for the installation. For example, to install a resistance welder, a millwright is needed to move the machine into position and fasten it to the floor; an electrician is required to make electrical connections; and a laborer may be needed to clean up the area before and after the machine is installed. For other installations, it may be necessary to consider ventilation, which would require the services of a sheet metal worker. Changes made to building foundations or the addition of water storage tanks would require carpenters and cement finishers.

Estimating the cost of installing a large number of machines can be done by classifying the machine according to weight. Table IV-2 lists typical costs, by trade or operation, used by one company in estimating machine costs. These values are representative and must be verified by the user.

Table IV-2. Machinery Installation Costs.

Machine Weight (lb.)	Costs by Trade or Operation				
	Rigging	Assembly	Setting	Leveling	Electrical
1,000 to 2,000	....	....	$ 50	$ 20	$125
2,000 to 4,000	....	....	100	50	$250 with disconnect and S. jack
4,000 to 8,000	$ 50	$200	150	100	$300 with starter; $500 without starter disconnect and swing
8,000 to 12,000	100	200 to 300	200	150	$300 to $500; $600 with MG set
Over 12,000	$200	$400	$300	$200	$700; $900 with MG set

Foundations:
   $30 per cu. yd. with reinforcing steel and anchor bolts
   $60 per cu. yd. with forms, reinforcing steel, and anchor bolts
   $10 per cu. yd. to excavate and haul away
   $2.50 per sq. ft. for saw cutting existing slab, break out, and haul away

Using these data in an example, the costs of installing a machine weighing 10,000 lbs. which is completely assembled upon arrival at the plant can be determined. The required operations include rigging, setting, leveling, and electrical installation. For an 8,000 to 12,000 lb. machine, these costs are $100, $200, $150, and $300, respectively, giving an estimated total cost of $750. This cost may be either slightly higher or lower than the actual cost but, on a project calling for the installation of several machines, the cost will average out to a reasonably accurate total cost.

If the facilities cost estimator has done the estimating, he will summarize his costs and forward them to the product cost estimating department where they will be prorated into the overall manufacturing cost estimate of the particular product being estimated. Major facility changes or improvements frequently affect more than one product and are prorated accordingly.

DIRECT MATERIAL COST

All materials on the bill of materials can be divided into two basic categories: (1) standard purchased parts, and (2) material for in-house fabricated parts.

Standard Purchased Parts

The estimator should obtain the cost of these parts from the purchasing department. His request should include an itemized list on a special form stating the size and quantity of each individual part needed. Any special requirements such as heat treatment or surface finish should be clearly indicated.

However, a formal request to purchasing sometimes requires more time than the cost estimate lead time will allow. To meet this situation, estimators often compile cost tables for standard purchased parts, *but such tables must be used cautiously*. Prices change frequently, and a wrong price can distort the entire estimate. In addition, purchasing is generally more aware of applicable price breaks at various purchasing volumes.

Purchasing should keep the estimating department informed of the latest cost figures on purchased parts. For companies using an integrated electronic data processing (EDP) system, such reports are easily obtainable.

Fabricated Parts

After requesting the cost of standard purchased parts from purchasing, the estimator draws up a list of every part to be made in-house which should include its quantity, size, weight, configuration, and type of raw materials required.

Raw Material Quantity. Next, the estimator analyzes each item to determine the exact amount of raw material or stock that must be purchased to manufacture the desired quantity. The cost of raw material for a piece is determined by multiplying the unit cost of the material by the weight of the rough stock used per piece. For instance, if a piece is machined, the amount of stock removed by machining must be added to the finished dimensions, and the volume is computed from these dimensions. If the piece is irregular in shape, it is divided into simple components, and the volumes of the components are computed and added together to give total volume. The volume is multiplied by the density of the material to obtain the weight. Experienced estimators are sometimes able to judge the weights of intricate pieces, such as castings, surprisingly close by comparing them with similar pieces.

Depending upon the type of stock used, estimators employ various formulas to determine material requirements and costs. In the case of a stamping fabricated from steel coil stock, the following four-step procedure can be used:

1) Weight of coil = gauge × width × density × length

2) Pieces per coil = $\dfrac{\text{length of coil}}{\text{length of multiple}}$

3) Weight per piece = $\dfrac{\text{weight of coil}}{\text{pieces per coil}}$

4) Piece material cost = piece weight × steel price per lb.

For bar stock, the length of the piece, plus facing and cutoff stock, is multiplied by the weight or price per inch of the diameter of stock as given in handbook tables.

The material that is lost in processing through scrapped pieces, butt ends, chips, etc., must be accounted for in an estimate. Losses vary from 1 to 12 per cent, depending upon the process, material, and practice. An average allowance of 5 per cent is often added to material estimates to distribute the bulk losses over the pieces produced.

A portion of this material loss is recoverable. Note the reclamation allowance in the final cost estimate shown in Fig. 4–1. Trimmings and cutoffs are usually sold as scrap, and, if so, the estimator can deduct an amount for scrap return from his material estimate.

COST ESTIMATE #3124	PART NUMBER 00046	PART NAME Name Plate Ass'y. for Customer	SHEET 1 OF 1 DATE 6/17/68

OPER. NO.	DEPT. NO.	DESCRIPTION	HOURLY PROD.	STD. TIME	SPEC. TOOLING	EQUIP.	REMARKS
1		Die Cast & Trim 2 cavity	250	.0040	11.000 6.000		
		Jobsetter & Stockhandler Allow.		.0007			
		Reclamation Allowance		.0006			
		Quality Control Allowance		.0003			
2		Load Plater Racks	600	.0017	1600		8 plater racks
3		CU-NI-CR Plate	16440	.0004			
		Stockhandler and relief		.0001			
4		Unload Plater Rack & Inspect	2349	.0004			
5		Assemble 2 Studs	300	.0033	6000		
		Jobsetter, Stockhandler, Acid Dip					
		& Strip		.0006			

MATERIAL SPEC.	S A E #903 Zinc				
COST/LB.	.1650	6 % LOSS	.0011	COMPILED	
WEIGHT EA.	.11#	FREIGHT		CHECKED	
LGTH CUTOFF		PAINT		SQ. IN.	
PIECE COST	.0182	PLATE	.0036	SQ. IN.	.0009
		TOTAL MTL		APP.	

TOTALS		.0121	24,600		STD. VOL.
@ /HR. 3.000	LABOR	.0362	STD. LABOR		
200.0% 100.0%	BURDEN	.0724 .0362	AMORT. TOOLING .0493		
	MATERIAL	0227 0100-	2 Studs		
	TOTAL PC. COST	.1775	PART NO.		

Fig. 4–1. Final cost estimate on a name plate assembly produced for a customer.

Price Breaks. The estimator and the purchasing department should be aware of applicable price breaks for quantity purchases of material. Some companies require estimators to use the price break applicable to the quantity of material that must be purchased for the part or product being estimated. Most companies, however, prefer to use the price break applicable to the company's actual purchase of raw material for *all* products. This is especially true for companies manufacturing a wide range of products fabricated from the same basic material. In this case, the estimating department must obtain information from purchasing regarding actual purchases.

Companies maintaining slitting and shearing equipment can buy mill stock and slit it to their required sizes. This permits them to purchase in greater quantity and reduces material costs. Estimators must add in the cost of slitting and shearing to arrive at the appropriate material cost for given products.

Material Cost Data. In general, the estimating department should know the costs of raw materials normally used for fabricating parts within the plant. Maintaining a file of current material costs reduces reliance upon purchasing and decreases estimating time.

Estimating Forms. To assist in computing material costs, an estimating form should be designed that is suitable for all of the most commonly manufactured items. As estimating situations arise that cannot be accommodated to the form, the form may be revised. For the sake of simplicity, as few forms as possible should be used. Estimating forms should provide sufficient description of the part so that any person familiar with that kind of manufacture will obtain a quick, clear picture of the item. (See Fig. 4–2.)

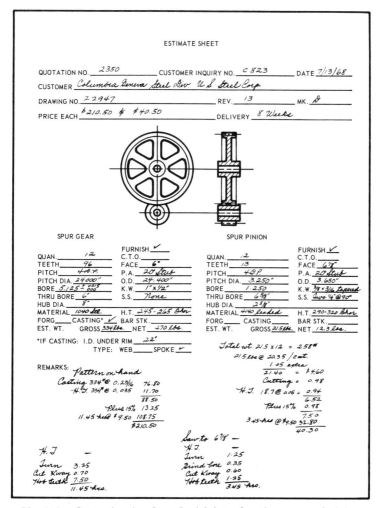

Fig. 4–2. Cost estimating form for job lots of mating gear and pinion.

Standard Costs. When standard costs are used for materials, the estimator, after computing raw material requirements, simply extends the figures by multiplying them by the standard cost per unit. Because standard costs are generally not established for materials purchased in small quantities, the estimator may have to use a combination of actual and standard costs.

Standard costs are especially useful in large job-shop plants manufacturing many items in varying production quantities. Over a given time period (such as a year) the accounting department establishes the average cost per unit of materials purchased by the company. This average, with appropriate allowances for price changes, serves as the standard cost for a given length of time.

Example. The form shown in Fig. 4–2 was designed for use in a general machine shop which has a department specializing in producing small job lots of gears. Space is allowed at the bottom of the sheet to indicate the operations required and estimated production time, in addition to dimensions and other pertinent data. Note the material calculations at bottom right on this estimate. (See also Fig. 3–6 for estimated material costs.)

TOOLING COSTS

An estimator must be well acquainted with tooling concepts. Tooling is an integral part of the manufacturing process, and new tooling and/or tooling changes are frequently required for new products. Therefore, the tooling cost estimate is necessary to determine the costs of alternative methods of tooling. Estimators must know their company's tooling capabilities and policies on new tooling costs, as well as have the ability to determine tooling requirements and costs for new products.

In addition to its role in the product cost estimate, tooling estimates are frequently requested by value analysis or value engineering to determine the cost differentials in proposed tooling changes on current production.

Because of the escalating costs of tools, tooling is one of the main items in a product cost estimate. More than ever before, tool costs are being treated as a distinct component of the product cost estimate, rather than being added to the estimate under the heading of factory burden.

Purchased Tooling

The tooling for most products is purchased from outside vendors; few manufacturing companies possess the specialized knowledge and equipment required to design and build their own tools. Therefore, tool cost is generally determined in much the same manner as the costs for any other purchased items, i.e., by using the purchase price of the tooling plus any applicable transportation, handling charges, and burden.

In-House Tooling

Estimating the costs involved when the company develops its own tools, in full or in part, is considerably more complicated. One large stamping plant has found the following percentages to be typical for tooling costs for metal stamping dies:

3%	Models and templates
12%	Designs
3%	Patterns, castings, and labor
50%	Build
22%	Material
10%	Tryout and testing
100%	Total tool cost

Estimating the cost of in-house tooling requires knowledge of previous tooling development costs. A record of previous cost data is very helpful.

Tool Cost Assignment

Once the cost of each tool has been determined (whether purchased or developed in-house), the estimator determines how many tools of each type are required for the desired production run. Information from previous production and data from vendors are useful in determining the life of perishable tools. In computing the number of perishable tools for a production run, allowance must be made for rework and scrapped parts. Historical data would provide useful information for this purpose also.

When durable tooling is usable only on the product being estimated, its entire cost (less any salvage value) must be assigned to the product. If the tooling can be used on similar products, the estimator must determine the extent to which it will be used and prorate it accordingly. Knowledge of future production plans will permit him to assign a reasonable percentage of the tool cost to the project being estimated. Because it is difficult to assign costs accurately for durable tooling, many companies include this item in overhead.

Data for Estimating Tooling Costs

Because tooling costs have become a major item, and because they are subject to frequent increase, estimators keep close check on tooling cost figures and frequently revise their recorded cost data.

Tool suppliers are the best source of cost information on tools. Past estimates may supply some guidance, but data taken from previous years must be upgraded to conform to present cost levels.

In purchasing tooling from suppliers, two or three bids should be requested for the same tooling in order to obtain a realistic idea of actual tooling costs.

By studying past cost figures for similar items of tooling, an estimator can derive a rate of cost increase for tools used by the plant. Then when a new product being estimated calls for tools similar to those used on a previous job, this percentage may be applied to the old tooling cost to arrive at a new tooling cost figure reflecting normal annual price increases.

MANUFACTURING TIME PERIODS

To develop labor costs and apply factory burden rates, the time required to accomplish each manufacturing step is necessary. Manufacturing operations can be separated into machine operations, machine setups, and process treatments.

Machine Operations

A machine operation is defined as all the steps necessary for a worker to complete a discrete procedure such as boring, cutting, or grinding a specific part.

Using the operation lineup compiled from the manufacturing processing plan, the estimator computes the time required for each operation, either by referring to standard data developed from previous products or by breaking each operation into its elemental motions.

When standard data on previous operations is not available, the estimator determines operation times by the use of predetermined time and motion elements. Handbook formulas can be used along with standard processing data tables giving operating information such as speeds and feeds, drilling and tapping times, etc., to compute these times.

The use of standard data compiled from actual plant processes is advisable since these data contain built-in allowances for such factors as machine downtime, operator fatigue, and maintenance. When predetermined time and motion data are used, the estimator must compute each item, thus introducing additional margin for error. The use of standard data also require less time than elemental motions, which can be an important factor when the workload is heavy or when estimate lead time is short.

Only a person with considerable background in the shop is competent to judge operation times with reasonable accuracy. Others should refer to recorded experience, such as cost-accounting department records, and compile a file of data on specific operations that have been performed in the past.

Competent estimators will also foresee work that an operator can perform while the machine is in cycle. Recognizing this additional available operator time and planning for it can reduce direct labor cost per piece.

Machine Setups

Setup time is estimated separately from machine operation time. It can be determined from standard data gathered from earlier products, or may be built up through a study of the elements or steps required to make the setup.

In job-shop operations it is difficult to use a formula to distribute setup costs. The owner-operator of a small shop will usually rely upon his past experience to establish a realistic figure.

For high-volume runs, however, setup costs are distributed on a unit basis by dividing total setup cost by the number of units in the production run.

Example:

$$\frac{\text{Number of Setups} \times \text{hrs/Setup}}{\text{Number of Units}} = \text{hrs/Unit}$$

Where:

Quantity $= 100,000$ Units

Number of Setups To Produce 100,000 Units $= 2$ Setups

Time Required To Perform One Setup $= 10$ hrs.

$$\frac{2 \times 10 \text{ hrs.}}{100,000 \text{ units}} = .00020 \text{ hrs.}$$

Therefore, in this example .00020 hours should be applied against each unit for setups.

This formula is applicable even if the production lots following individual setups vary in quantity. The theory is that, since a certain number of setups are required to produce a full production run, the total setup cost should be applied evenly to each item in the production run. When lot sizes are equal in quantity, setup costs assignable to each unit would be the same whether the costs were assigned on a total production run basis or on the basis of separate production lots. If the latter costs are needed, then setup costs must be computed on the basis of individual production lots.

Process Treatments

Time computations may not be necessary for process treatments such as plating, painting, and dipping. Instead of time, material usage frequently serves as the basis for labor and factory burden cost assignment.

Additional material is not added to the product with process treatments such as cleaning, curing, sand blasting, and drying. For these processes, time should be computed as for machine operations.

DIRECT LABOR COSTS

By this time the estimator has computed direct material, tooling, and facilities costs, and has determined manufacturing operation times. He should now be able to visualize the labor required to conduct the manufacturing processes. Labor costs are derived by extending the times for component manufacturing operations, and by applying appropriate labor cost rates.

The accounting department furnishes these rates in most companies. The rates generally include the company's contributions to social security, group insurance, and retirement plans, in addition to hourly wages. Some companies add an overhead factor to cover other activities related to the personnel function.

FACTORY BURDEN

Factory burden includes all costs incurred by the company that cannot be traced directly to specific products. The accounting department determines burden rates, and the estimator assigns burden costs to individual items on a formula basis.

Fixed and Variable Burden

Burden consists of fixed and variable categories, and separate rates are often established for each. Fixed burden includes all continuing costs regardless of the production volume for a given item, such as salaries, building rent or mortgage payments, and insurance. Variable burden costs, on the other hand, increase or decrease as the volume of production rises or falls. Indirect material, indirect labor, electricity used to operate equipment, and certain tooling are classified as variable burden items.

Cost Centers

Variable burden may be assigned more accurately by using different rates for cost centers within the plant.

A cleaning and plating operation, for example, may be set up as a cost center. In establishing the estimate for a chromium plated automobile bumper, the estimator would determine the direct material cost of the chromium plating material, and then apply the cleaning and plating burden by using the cost center burden rate.

For example, if the cleaning and plating cost center burden rate was established as 170 per cent of direct material, and the direct material cost was $.230, the cleaning and plating burden would be $.391. Total cleaning and plating costs (the sum of direct material and cost center burden) would be $.621.

When separate cost centers are used for variable burden, the estimator arrives at a total product cost, exclusive of variable burden, before calculating fixed burden. The estimate for the automobile bumper could be set up as shown in Table IV-3.

Table IV-3. Cost Estimate for Automobile Bumper.

Cost Element	Cost per Part	Burden Rates (%)		Burden
		Fixed	Variable*	
Direct material steel (8 lbs.@ $7.25/hwt.)	$ .580	....	....	....
Chrome plate 1.2 lbs.@ $19.76/hwt	.230	....	170	$ .391
Direct labor stamping .16 hrs.@ $7.15	1.144	....	310	3.546
Tooling	.780	....	210	1.638
Total direct costs	$2.734	73	....	1.996
Total burden				$7.571

*Note that different factors are used to compute variable burden. Direct labor is used for the stamping operation, while direct material is used for cleaning and plating. Total cost for the bumper would be $10.305, the sum of total direct costs and total burden.

Burden Assignment Methods

The method of assigning factory burden to individual products differs from industry to industry, and even from one company to another within an industry. Any quantifiable product factor may serve as the basis for assignment of factory burden as long as consistent use of the factor across the entire product line results in full and equitable burden distribution. The factors most frequently used are:

1) Direct labor cost
2) Direct material cost
3) Number of parts produced.

One large manufacturer assigns both fixed and variable burden on the basis of direct labor cost. Fixed burden is assigned at 100 per cent of direct labor, and variable burden at 340 per cent. Thus, the burden cost on an item requiring $.04 in direct labor is $.176−$.04 for fixed burden and $.136 for variable burden.

Depending upon a company's product line, direct material cost or number of parts produced may be a more accurate factor for distribution of burden costs. Direct material cost is favored by process-type industries and other companies manufacturing products in which direct material is the major cost element. A company that makes a stable line of products selling for approximately the same price can use the number of items produced to assign burden.

Burden costing for any manufacturing operation may vary in its use of methods and assignment factors. What is important is that the procedures chosen furnish costs for the parts and products the company manufactures.

It is not necessary for the estimator to know how to set up an accounting system for burden assignment, but he must understand the existing system in sufficient detail to know what costs are covered by burden rates and which costs must be computed separately for each item required to be estimated.

TOTAL MANUFACTURING COST

The total product manufacturing cost is the sum of the cost of in-house manufactured parts, purchased items, subassembly costs, and final assembly costs. The total manufacturing cost for a part is determined by adding facilities, direct material, tooling and equipment, direct labor, and burden costs.

Tooling, Equipment, and Facilities Costs

Tooling, equipment, and facilities costs applicable to specific parts are sometimes included as integral elements of the manufacturing cost of the item and are prorated over a given number of production units before factory burden is determined. Alternatively, these costs may be added to the finished product estimate as a separate item and then prorated over the number of units produced. Customer specifications or company practice usually indicate which method is to be used.

Applying Cost Rates

In some companies the estimating department's responsibility for total manufacturing cost ends with establishing direct material, tooling, equipment, facilities, and labor costs. The merits of having accounting add wage and burden rates from applicable cost centers and extend the figures depend on the organization of the particular company.

SELLING PRICE

Once the accounting department or estimating department has applied burden rates and developed a total manufacturing cost, a percentage for profit must be added to arrive at the selling price. Because of the sales or marketing department's acquaintance with customers and markets, the responsibility for this calculation is logically theirs, although it may sometimes be made by the estimating or accounting departments.

The marketing department normally controls the entire pricing structure and calculates price breaks at various purchasing volumes. This enables them to establish a realistic range of price breaks to ensure overall profit. In addition, marketing is usually given the prerogative of varying the profit margin to meet the circumstances of the sale. For instance, in order to stimulate sales, the marketing department may lower the profit margin in order to saturate the market. Modern marketing departments use a vast array of quantitative and qualitative techniques, such as market surveys, opinion samples, and test markets, to measure market demand and to calculate which profit margins and sales levels will result in maximum overall profit.

REFERENCES

1. C. Northcote Parkinson, *Parkinson's Law* (Boston: Houghton Mifflin Co., 1957).

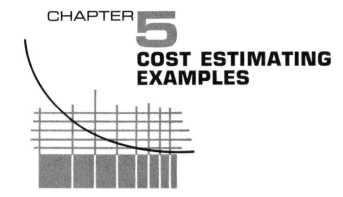

CHAPTER 5
COST ESTIMATING EXAMPLES

This chapter gives examples of cost estimates that illustrate the principles and procedures of estimating described in Chapter 4.

The estimating forms illustrated in this chapter and in previous chapters are examples of those used in industry, and their format can be employed as shown, or adapted to fit individual circumstances. The cost figures applied in individual examples are approximations only and should not be used as actual cost data.

ESTIMATING DIE CASTING MACHINING COSTS

Assume that the cost estimating department has been requested to prepare an estimate for a cam drive bracket. Detailed process sheets and tool design drawings (not shown) accompanied the cost request, and all necessary fixtures, gages, and other tools have been designed and are listed on the manufacturing process plan. Fig. 5–1 shows the assembly drawing, with machining dimensions, of the part being estimated.

Each individual manufacturing operation listed on the manufacturing process plan will be analyzed, and the material, labor, burden, and tool costs per unit summarized on the cost estimate form (Fig. 5–2).

Purchased Parts

The bracket (Detail 1 in Fig. 5–1), an aluminum die casting ASTM B 85-49 alloy SC-6, is to be purchased without any machining. The purchasing department requested cost quotations for the die casting from several companies. The lowest quoted cost was $22 per 100 pieces ($.220 each) plus a die cost of $1,810. The bushings, Detail 2, are standard sintered metal powder parts purchased at a cost of $.038 each. The stud, Detail 3, will be purchased at a cost of $.037 each.

51

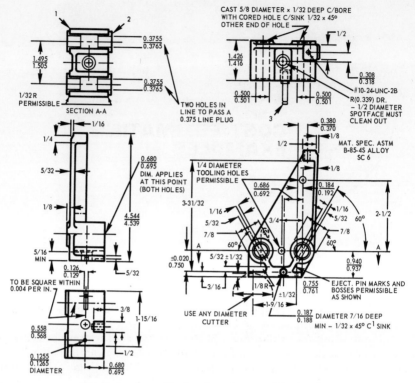

Fig. 5-1. Assembly drawing of a cam drive bracket.

These material and tool costs are entered on the cost estimate form shown in Fig. 5-2. In Operation 001, the $.220 bracket is entered in the "Material Cost" column, and the $1,810 (die cost) is entered under the "New Tool Description." The cost of the stud ($.037) is entered as the material cost for Operation 100, and the bushings ($.038 each) as material cost for Operations 110 and 120.

Handling and Machine Times

The time required for each separate manufacturing operation must be computed before the appropriate labor, burden, and tooling costs can be summarized on the estimate form (Fig. 5-2).

Standard Data. Standard data compiled from time studies of previous plant operations were used for estimating the handling and machine times in this estimate. Table V-1 gives examples of these data.

Hole Reaming. Hole Y is the ¼ in. diameter tooling hole on the centerline of the two bushing holes (see Fig. 5-1). Operation 010 calls for this hole to be reamed on machine 07-015. This machine number code indicates that the operation is to be performed on a single-spindle drilling machine in the drilling department (No. 07).

PART NAME *Bracket, cam drive*   PART NO *790-3X*
ENG MEMO NO_____SHEET _/_ OF _/_ LOT SIZE_____ DATE *6/17/68*
LABOR $ *0.536*   MAT'L $ *0.409*   TOTAL L&M $ *0.945*
TOOL DESIGN $ *547*   TOOLROOM $ *2602*   DIES $ *1810*   TOTAL TOOLS $ *4959*

OPER	DEPT	LC	OPER OR MAT'L DESCRIPTION	MAT'L QUAN	MAT'L COST	P C	LABOR STD	L G C	L&B COST	NEW TOOL DESCRIPTION	TOOL COST
001	30	R	*Purchase complete*								
			to B/P except machining								
			& Assy. (1) BZ-790-3	1.000	$ .220					*Die Cost ($1810)*	
010	07	05	*Ream "y"*			E	.149	6	$ .012	*No new tools*	
020	06	10	*Straddle Mill H & E ²⁵*			E	.550	6	$ .050	*Milling Fixture*	275⁰⁰
										(2) Side milling cutters	125⁰⁰
										Flush Pin 6 A	125⁰⁰
										Snap Gage	20⁰⁰
030	18	51	*Burr H & E*			E	.430	7	.030	*Standard Tools*	
040	14	08	*Bore B & U & chmfr.*			E	1.584	5	.158	*Boring Fixture*	420⁰⁰
										Set Master	45⁰⁰
										(2) Boring Bars	81⁰⁰
										(4) Tube	
050	07		*Drill - C'Bore - Drill etc.*			E	2.091	6	.167	*Drill Jig*	358⁰⁰
										.339 C'Bore	13⁰⁰
										½ C'Bore	13⁰⁰
										C'Bore Fixture	26⁰⁰
060	06	09	*Mill - M-L-S ¹²⁵*			E	.299	6	.027	*Fixture*	214⁰⁰
										Cutter	20⁰⁰
										Flush Pin Gage	51⁰⁰
										Centrality Ga & Tuler	381⁰⁰
070	01	19	*Wash*				—				
080	09	07	*Burr Complete*			E	.201	9	.014	*No new tools*	
090	32	N	*Inspect*				—	—	—		
100	21	A03	*Assemble - AK -*				.195		.014		
			790-125	1.000	.037					*Fixture*	46⁰⁰
110	21A	03	*Assemble AE & AF*				.460		.032	*Fixture & Studs*	350⁰⁰
			BU-675	2.000	.076						
120	21A	02	*Assemble AA & AD*				.460		.032	*Studs*	27⁰⁰
			BU-675	2.000	.076						
130	21	AN	*Inspect Final*								
					$ .409				$ .536		$ 260⁰⁰
										Die Cost	1810
										Tool Design	547
										Total tools $	4959

Fig. 5-2. Cost estimate for the bracket assembly.

Time Study Operation Sheet. The time study operation sheet for Operation 010 is shown in Fig. 5-3. The data entered on this sheet were taken from Tables V-1 and V-2. Assuming that the chips are brushed from the machine table every third piece, the time required for Work Act No. 4 (see Fig. 5-3) is divided by 3; thus a time of .010 min. per piece is entered on the sheet. The total floor-to-floor time is estimated as .149 min.

Labor and Burden. The labor and burden rate per minute for department No. 07 is $.08. The cost per piece is .149 min. × $.08/min = $.012 which is entered in the "L & B (labor and burden) Cost" column of the estimate form (Fig. 5-2) for Operation 010.

Table V-1. Standard Data for Time Study.

Work Act Description	Weight (lbs.)	Standard Time (min.)
Tote box to machine — counting	.00– 1.00	.030
	1.01– 2.00	.035
	2.01– 3.00	.045
	3.01– 4.50	.050
Piece in and out of jig — place	.00– .50	.099
	.50– 3.00	.126
	3.00– 5.00	.157
Piece in and out of jig — locate	.00– .50	.132
	.51– 3.00	.168
	3.01– 5.00	.193
Piece on and off block or nest	.00– 1.00	.045
	1.01– 2.00	.079
	2.01– 3.00	.103
Tighten and loosen handwheel — all types spin on and off	. . . .	.079
Open and close hinged jig leaf:		
one	. . . .	.043
two	. . . .	.064
Tighten and loosen — T wrench:		
one nut	. . . .	.189
two nuts	. . . .	.285
Tighten and loosen thumbscrew	. . . .	.096
Open and close sliding clamp	. . . .	.040
Up and down spindle:		
full travel	. . . .	.067
normal travel	. . . .	.041
less than ½ in. travel	. . . .	.020
Change tool — quick change chuck	. . . .	.103
Oil tap with brush or up can	. . . .	.049
Place and remove bushing — each	. . . .	.069
Wash jig with soda water	. . . .	.070
Wash jig with nozzle	. . . .	.070
Brush off table	. . . .	.189
Gaging:		
threaded gage go — no go	.00– .37	.250
	.38– .74	.380
scale	. . . .	.110
feeler gage	. . . .	.110
snap gage	. . . .	.110
gage — counter sink with screw	. . . .	.125
Jig handling elements:		
up to first spindle and position	.00– 4.00	.040
	4.01–25.00	.066
	26.00–40.00	.086
Shift — hole to hole	.00– 4.00	.020
	4.01–25.00	.030
	26.00–40.00	.050
Away from last spindle	.00– 4.00	.030
	4.01–25.00	.046
	26.00–40.00	.058

TIME STUDY OPERATION SHEET

DEPT. _07-015._   DATE _6/17/68_   MAT'L DES _Alum. Die Casting_
PART NO. _790-3X_   NAME _Bracket cam-drive_
OPER. NO. _010_   TYPE _Ream_
MACHINE NO. _____   TYPE _Delta Drill Press_
TOOLS _Standard holding block_ SPEED _____   FEED _Hand_

NO.	DESCRIPTION OF WORK ACT.	OCCURRENCE	STD. TIME, MIN.	TIME ALLOWED, MIN.
1	Piece from tote box to machine and count (0 to 1 lb.)	1:1	0.030	0.030
2	Piece on and off block or rest	1:1	0.045	0.045
3	Up & down spindle (less than 1/2" travel)	1:1	0.020	0.020
4	Clean block, 3 places	1:3	0.030	0.010
5	Plug gage	1:20	0.010	0.005
			Total handling time	0.110
	Machining time			
6	Ream (0.2505)	0.140 in @	0.105 in./min.	0.019
		overtravel	0.015	0.015
				0.032
	From Table V-2 min. time allowance is			0.039
			Total time	0.149

Fig. 5–3. Time study operation sheet for Operation 010.

Straddle Milling. Operation 020 calls for straddle milling surfaces H and E. The process sheets stipulate that the part is to be milled in a fixture by 8-in.-diameter side milling cutters on a milling machine. The spindle speeds and table feeds for this machine are given in Table V-3. The estimator notes that a small duplex mill would be a better machine to use. However, this machine is loaded to capacity and is not available.

Handling Time. For this operation, standard data are used to estimate the handling time only. The work acts are defined in a manner similar to that used for Operation 010. The estimated handling time is .275 min.

Machine Time. The machine time is estimated as follows. Using a recommended 800 sfpm., the required rpm. of the spindle is determined by sfpm/cutter perimeter. Thus, $800/\pi \times 8/12 = 384$ rpm. From Table V-3, the nearest spindle speed for the machine is 380 rpm. A feed of .005 in. per tooth is selected for the operation. A cutting tool catalog reveals that an 8-in.-diameter HSS side milling cutter has 26 teeth. With these data, the appropriate formula is

Table V-2. Standard Data for Drilling, Reaming, and Tapping Aluminum.

		Drilling	Tapping	
Diameter of Tool (in.)	Approach of Drill Point (min.)	Standard Time (min/in)*	Tap Size	Standard Time (min/in)*‡
$3/32$	.028	.080	2–56	.09
$1/8$	.037	.090	3–48 and 4–48	.08
$5/32$	.047	.100	4–40	.07
$3/16$	.056	.115	5–44	.08
$7/32$	.066	.125	5–40	.07
$1/4$	.075	.130	6–40	.08
$9/32$	.084	.135	6–32	.06
$5/16$	.094	.140	8–36	.08
$11/32$	.103	.145	8–32	.07
$3/8$	.113	.150	10–32	.09
$13/32$	.122	.155	10–24	.07
$7/16$	.131	.160	12–28	.09
$15/32$	.141	.165	12–24	.08
$1/2$	.150	.170	$1/4$–28	.10
$17/32$	.160	.175	$1/4$–20	.07
$9/16$	.169	.180	$5/16$–24	.11
$5/8$	.188	.185	$5/16$–18	.08
			$3/8$–24	.14
			$3/8$–16	.09
			$7/16$–20	.13
	Reaming*†			
$3/32 - 9/64$	....	.075	$7/16$–14	.09
$5/32 - 13/64$	....	.090	$1/2$–20	.15
$7/32 - 17/64$	....	.105	$1/2$–13	.10
$9/32 - 3/8$	....	.120		
$13/32 - 1/2$	....	.135		

*Use .039 standard minute as a minimum time.
†Add a constant of .015 standard minute to the reaming times for overtravel of the reamer.
‡Add the tap diameter to the length of the tapped hole to obtain total tool travel.

Table V-3. Data for Milling Machine.*

Spindle Speeds (rpm.)				Table Feeds (ipr.)			
50	126	302	760	$3/4$	2	$5^3/8$	14
63	159	380	955	1	$2^1/2$	7	18
79	200	475	1200	$1^1/4$	$3^1/4$	$8^7/8$	23
100	250	600	1500	$1^5/8$	$4^1/4$	11	30

*Machine and tool manufacturers differ regarding optimum speeds and feeds; the ones shown here are illustrative only and are not recommendations.

used to determine the optimum feed in inches per minute for the operation:

$$\text{rpm.} \times \text{Number of Teeth} \times \text{Feed per Tooth} = \text{Feed (in/min)}$$

$$380 \times 26 \times .005 = 49.4 \text{ in/min}$$

Referring to Table V-3, the maximum feed rate is 30 in/min. The length of cut is $2^3/8$ in. plus $1/8$ in. overtravel, giving a total of $2^1/2$ in.

Next, an allowance must be made for *cutter approach* (the distance the work must travel into the cutter before the cutter is at the full depth of cut):

$$\text{Approach} = \sqrt{r^2 - (r - c)^2}$$

Where: c = depth of cut (in.)
r = cutter radius (in.)

Using a depth of cut of 1.4 in., the approach is calculated to be 3.04 in. With an approximate length of cut of 2.50 in. plus a 3.04 in. approach, the table travel is 5.54 in. The machine time is calculated by dividing the table travel by the feed per inches. Thus 5.54 in. ÷ 30 in/min = .185 min.

An allowance must also be made for rapid *machine table traverse*. In this case .030 min. is needed for rapid approach and .060 min. for rapid return. Adding the cutting time of .185 min. and the rapid traverse time of .090 min. gives a total time of .275 min.

Total Time. The total man-machine time for Operation 020 is .275 min. handling time plus .275 min. machine time, or .550 min. per piece.

Labor and Burden. The labor and burden rate for the milling department is $.09 per minute. Thus, the cost per piece is .550 min. × $.09 per min. = $.050. These costs are entered on the estimate form (Fig. 5–2).

Summary of Operations. The remaining operations, because of their number, will be summarized in Table V-4 rather than presented in detail. The cost for each operation is entered on the estimating form (Fig. 5–2).

Table V-4. Summary of In-Plant Manufacturing Operations and Their Costs.

Operation Number	Description	Time (min.)	Multiplied by L & B ($)	Estimated Cost
001	purchased item	. . . .	. . . .	. . . .
030	burr H and E	.430	.07	.0301
040	bore G and U, chamfer AG and B	1.584	.10	.1584
050	drill, counter bore, etc.	2.091	.08	.16728
060	mill M-L-S	.299	.09	.02691
070	wash*	. . . .	. . . .	no charge*
080	burr complete	.201	.07	.01407
090	inspect*	. . . .	. . . .	no charge*
100	assemble AK	.195	.07	.01365
110	assemble AE and AF	.460	.07	.03220
120	assemble AA and AC	.460	.07	.03220
130	final inspection*	. . . .	. . . .	no charge*

*In this company wash department, inspection, trucking, receiving, and shipping costs are considered as overhead. This practice is not universal and varies greatly from company to company.

Total Manufacturing Cost

When all the manufacturing operations are entered on the estimate form (Fig. 5–2) with their corresponding costs, these costs are added to arrive at a material cost per unit, a labor and burden cost per unit, and a total tooling cost.

The total cost of manufacturing the cam drive bracket in Fig. 5–1 can be summarized as follows:

Material, cost/unit	$ .409
Labor and burden cost/unit	+.536
Total material and L & B/unit	$ .945
Tooling:	
Die cast die	$1,810
Fixtures, cutting tools, and gages	2,602
Tool design	+547
Total tooling cost	$ 4,959

In order to assign the tooling cost to individual units, the production quantity must be known. Assuming a lot size of 2,000 units, the tooling cost per unit is calculated as follows:

Material, labor, and burden per unit	$ .945
Lot size	× 2000
Total material, labor, and burden	$ 1,890
Total tooling cost	+4,959
Total costs	$ 6,849

When the total cost of $6,849 is divided by the number of units produced (2,000) the total cost per unit is $3.42. The total cost per unit declines as the production quantity increases and tooling cost is absorbed by more units. Total cost per unit for a production quantity of 3,000, for example, would be $2.60; for 5,000 units, the unit cost decreases to $1.94.

Overhead and Profit

Allowances are made for general and administrative expenses (overhead) and profit before a price is quoted to the customer. The estimating department may make these calculations based on predetermined rates, but in most companies the estimating department forwards the cost estimate to either the accounting or sales department to determine the final price.

ESTIMATING MACHINING COSTS FOR AN ALUMINUM FORGING

Assume that a customer has requested a price quotation on the manufacture and delivery of 2,000 completed forged aluminum housings. An engineering drawing (Fig. 5–4) and complete blueprint (not shown) accompanied the request. The customer's schedule specifies initial delivery of 500 parts within eight months, with delivery of the remaining 1,500 parts in three lots at three-month intervals. Because the company receiving the cost request does not have forging capabilities, it purchases rough forgings on orders of this type and machines them to specifications.

In order to give an accurate price quotation, the company prepares a complete cost estimate. The steps followed are: (1) preliminary screening, (2) gath-

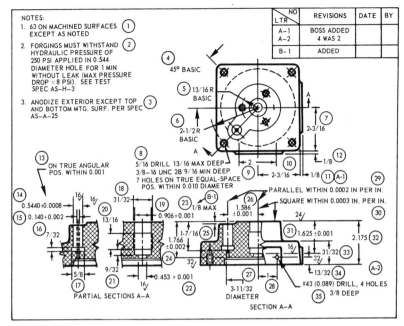

Fig. 5–4. Forged aluminum housing.

ering information, (3) compiling the data, (4) extending the data by adding costs, and (5) establishing the selling price.

Preliminary Screening

The sales department studies the request in terms of profitability, plant capabilities, and scheduling to determine whether to submit a quotation. The request meets these criteria, and the sales department sends a cost estimate request to the estimating department, specifying an estimate deadline.

Gathering Information

Upon receiving the request, the estimating department analyzes it for necessary information, and requests certain data from other departments. These departments are:

(1) The tool and process engineering departments, which have knowledge of practical methods, proved equipment, and realistic machine times, and which determine the manufacturing processes and tooling necessary to produce the part.

(2) The purchasing department which obtains prices on forgings and other items listed for outside purchase.

(3) The production control department which advises the estimating department whether the customer's specified delivery dates can be met without upsetting existing production schedules. This information can be furnished only after the manufacturing plan has been established and machine times calculated.

Compiling the Data

As the groups listed above forward information, the estimator enters the data on a cost estimate form.

The estimate is prepared in duplicate: a "reference" copy containing all information except costs, and an "action" copy on which cost data are entered. The reference copy is available to any interested personnel, but the action copy is distributed on a "need-to-know" basis only.

The Estimate Form. Fig. 5–5 shows the first page of the cost estimate. At this point in the preparation of the estimate, however, only the operation number and description, machine name, list of tools, and the notation of standard tools would be entered on the form.

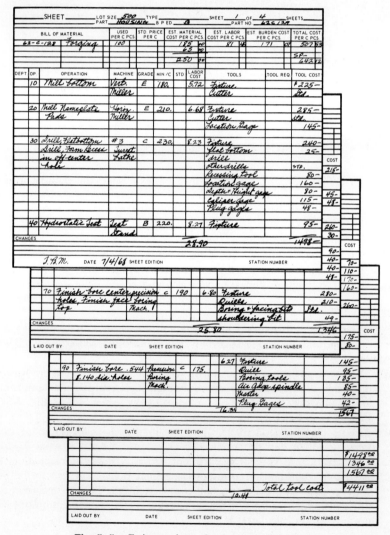

Fig. 5–5. Estimate sheets for forging of Fig. 5–4.

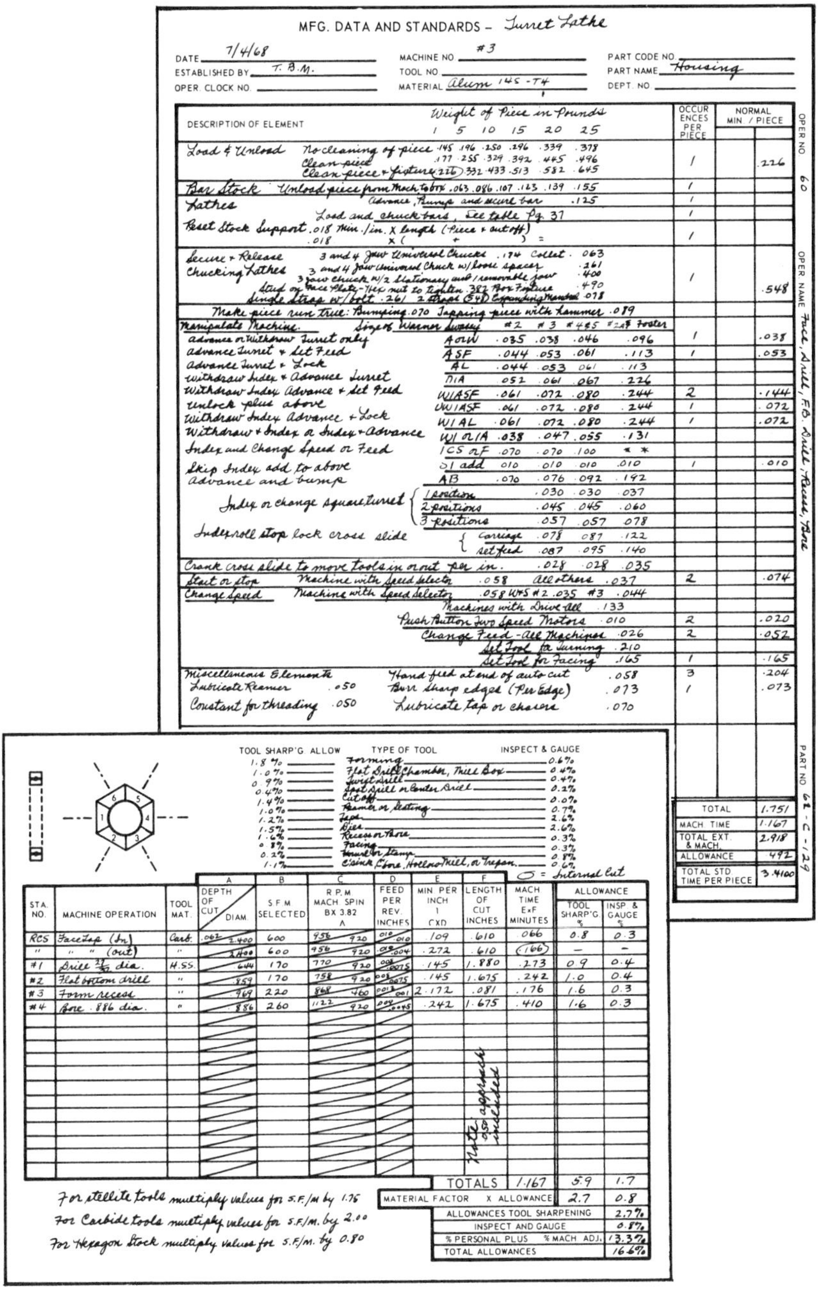

MFG. DATA AND STANDARDS – Turret Lathe

DATE 7/4/68 — MACHINE NO. #3 — PART CODE NO.
ESTABLISHED BY T. B. M. — TOOL NO. — PART NAME Housing
OPER. CLOCK NO. — MATERIAL Alum 14S-T4 — DEPT. NO.

DESCRIPTION OF ELEMENT	Weight of Piece in Pounds: 1 5 10 15 20 25	OCCURENCES PER PIECE	NORMAL MIN. / PIECE
Load & Unload	No cleaning of piece .145 .196 .250 .296 .339 .378; Clean piece .177 .255 .329 .392 .445 .496; Clean piece + fixture (.226) .332 .433 .513 .582 .645	1	.226
Bar Stock Lathes	Unload piece from Mach to box .063 .086 .107 .123 .139 .155	1	
	Advance, Bump and secure bar .125	1	
	Load and chuck bars, See table Pg. 37	1	
Reset Stock Support	.018 Min./in. X length (Piece + cut off) .018 x (+) =	1	
Secure + Release Chucking Lathes	3 and 4 Jaw Universal Chucks .174 Collet .063; 3 and 4 Jaw Universal Chuck w/loose spacer .261; 3 jaw chuck w/2 stationary and 1 removable jaw .400; Stud or Face Plate – Hex nut to tighten .382 Box Fixture .490; Single Strap w/bolt .261 2 straps (.54) Expanding Mandrel .078	1	.548
	Make piece run true: Bumping .070 Tapping piece with hammer .089		
Manipulate Machine	Sizes Warner Swasey #2 #3 #4&5 #2A Foster		
advance or withdraw Turret only	A or W .035 .038 .046 .096	1	.038
advance Turret + Set Feed	ASF .044 .053 .061 .113	1	.053
advance Turret + Lock	AL .044 .053 .061 .113		
withdraw Index + Advance Turret	WIA .052 .061 .067 .226		
Withdraw Index Advance + Set Feed unlock plus above	WIASF .061 .072 .080 .244	2	.144
	UWIASF .061 .072 .080 .244	1	.072
Withdraw Index Advance + Lock	WIAL .061 .072 .080 .244	1	.072
Withdraw + Index or Index + Advance	WI or IA .038 .047 .055 .131		
Index and Change Speed or Feed	ICS or F .070 .070 .100 * *		
Skip Index add to above	SI add .010 .010 .010 .010	1	.010
advance and bump	AB .070 .076 .092 .192		
Index or change Square Turret	1 position .030 .030 .037; 2 positions .045 .045 .060; 3 positions .057 .057 .078		
Index roll stop lock cross slide	Carriage .078 .087 .122; set feed .087 .095 .140		
Crank cross slide to move tools in or out per in.	.028 .028 .035		
Start or stop	Machine with speed selector .058 All others .037	2	.074
Change Speed	Machine with speed selector .058 W+S #2 .035 #3 .044		
	Machines with Drive All .133		
	Push Button Two Speed Motors .010	2	.020
	Change Feed – All Machines .026	2	.052
	Set Tool for Turning .210		
	Set Tool for Facing .165	1	.165
Miscellaneous Elements	Hand feed at end of auto cut .058	3	.204
Lubricate Reamer .050	Burr sharp edges (Per Edge) .073	1	.073
Constant for threading .050	Lubricate tap or chasers .070		

OPER NO 60
OPER NAME Face, Drill, F.B. Drill, Recess, Bore
PART NO. 68-C-129

TOTAL	1.751
MACH TIME	1.167
TOTAL EXT. & MACH.	2.918
ALLOWANCE	.492
TOTAL STD TIME PER PIECE	3.4100

TOOL SHARP'G ALLOW	TYPE OF TOOL	INSPECT & GAUGE
1.8 %	Forming	0.6 %
1.0 %	Flat Drill, Chamber, Mill Box	0.4 %
0.9 %	Twist Drill	0.4 %
0.4 %	Spot Drill or Center Drill	0.2 %
1.4 %	Cut off	0.0 %
1.0 %	Reamer or Sizing	0.7 %
1.2 %	Taps	2.6 %
1.5 %	Dies	2.6 %
1.6 %	Recess or Bore	0.3 %
0.8 %	Facing	0.3 %
0.2 %	Knurl or Stamp	0.8 %
1.1 %	C'sink C'bore, Hollow Mill, or Trepan	0.6 %

○ = Internal Cut

STA. NO.	MACHINE OPERATION	TOOL MAT.	A DEPTH OF CUT / DIAM	B S.F.M. SELECTED	C R.P.M. MACH. SPIN BX 3.82 / A	D FEED PER REV. INCHES	E MIN PER INCH 1 / CXD	F LENGTH OF CUT INCHES	MACH TIME ExF MINUTES	ALLOWANCE TOOL SHARP'G %	ALLOWANCE INSP & GAUGE %
RCS	Face Top (In)	Carb.	.062 / 2.400	600	956 / 920	.010 / .010	.109	.610	.066	0.8	0.3
"	" " (out)	"	2.400	600	956 / 920	.018 / .004	.272	.610	(.166)	–	–
#1	Drill 41/64 dia.	H.S.S.	.644	170	770 / 920	.008 / .0075	.145	1.880	.273	0.9	0.4
#2	Flat bottom drill	"	.859	170	758 / 920	.008 / .0075	.145	1.675	.242	1.0	0.4
#3	Form recess	"	.969	220	868 / 460	.0013 / .001	2.172	.081	.176	1.6	0.3
#4	Bore .886 dia.	"	.886	260	1122 / 920	.004 / .0045	.242	1.675	.410	1.6	0.3

Note applicable .050 included

TOTALS	1.167	5.9	1.7
MATERIAL FACTOR X ALLOWANCE		2.7	0.8
ALLOWANCES TOOL SHARPENING			2.7 %
INSPECT AND GAUGE			0.8 %
% PERSONAL PLUS % MACH. ADJ.			13.3 %
TOTAL ALLOWANCES			16.6 %

For stellite tools multiply values for S.F./M by 1.75
For Carbide tools multiply values for S.F./M. by 2.00
For Hexagon Stock multiply values for S.F./M. by 0.80

Fig. 5–6. Standard data for turret lathe.

Labor and Machine Costs. The labor grades and machine times, based on standard data collected and maintained by the estimating department, are entered in their respective columns. Fig. 5–6 shows a typical form for applying standard data to a turret lathe for Operation 60. On the front side of the form (Fig. 5–6 *top*) the estimator lists all the machine manipulations normally encountered in the appropriate columns. The circled items are those relating to the component affected by this operation.

Variable Operations. The reverse side, Fig. 5–6 *bottom*, contains space for calculating the time for variables, plus applicable standard data. This side is completed first, starting with the entries for tool station number, machine operation, and tool material. Diameters of cuts are entered in column A. Values shown in surface feet per minute (sfpm.) in column B are based on standard data for the pertinent material available from handbooks. In column C, the left-hand figures are derived from the formula at the head of the column, and those to the right are machine speeds for the applicable machine. Similarly, the left-hand figures in column D represent standard suggested feeds, while those to the right are actual machine feeds.

The values for "length of cut" include an arbitrary amount for "approach," since the operator cannot be expected to advance the tool by hand and engage the automatic feed exactly at the point of contact with the work. In the "Machine Time" column, internal (concurrent) cuts are circled to indicate that they are not to be included in the total machine time for the piece.

Tool Sharpening and In-Process Gaging. Percentage allowances for tool sharpening and in-process gaging are found in the standard data table at the top of the sheet, and are entered in their respective columns. These figures are based on AISI B1113 material and, like all other standard data used in this estimate, have been found by the company through its studies to represent a norm for its particular operations.

Adjustment of Allowance Figures. After totals for the last three columns are entered, the allowance figures are adjusted to suit the material being cut. The forging's free-machining aluminum wears cutting edges 55 per cent less than AISI B1113, with a proportionate reduction in the need for tool inspection, resulting in the reduced allowances shown. These are added to constant personal and machine adjustment values to obtain a percentage value for all allowances.

Completing the Form. On the front side of the form, all applicable units of the tabulated elements are encircled, the number of their occurrences are entered, and the two are multiplied. The results, representing the actual accountable time for these elements, are entered in the last column and added. To this sum is added the total machine time, entered from the reverse side, giving a value for total machine manipulation and machining time. Multiplying this by the total allowance percentage gives the allowance value in minutes, which is added to the last previous value to arrive at the total time per piece for the operation.

Cost Extensions

When all job grades and machine times have been entered, the reference copy of the estimate is retained by the estimator, and the action copy is forwarded to the cost department (separate from the estimating department in this company). The cost department enters cost extensions, computing them on the basis of job grades and machine times established by the estimating department. A summary of the extensions, detailed in the following sections, is given in Table V-5.

Table V-5. Derivation of Selling Price for Aluminum Forgings in Lots of 100.

Materials
    Blank forgings @$1.85 × 100 = $185
    Anodizing = $65/100 pieces +   65
         Total Material Cost $250 $250.00

Labor
    Direct labor for all machining and ancillary operations $ 81.46   81.46
    Standard burden rate   × 210%
                        Burden Cost $171.07   171.07
    MLB (Sum of Materials, Labor and Burden) $502.53

General and Administrative Factor (15%)
                      MLB $502.53
                           ×.15
                      $ 75.38   75.38
                            Total Cost $577.91

$$\text{Profit Percentage} \left(\text{Selling Price} = \frac{\text{Total Cost}}{90\% \ (100\% - 10\% \ \text{profit})} \right)$$

$$\frac{\$577.91}{.90} = \$642.12$$

Labor. Typical rates used for eight classes of labor are shown in Table V-6. To facilitate calculation, these rates are for 100 min. rather than per hour. Thus, the direct labor cost for Operation 10, for example, is found by multiplying the minutes of machine time (180) by the Class E rate ($3.18) to arrive at a cost of $5.72 per 100 pieces.

Table V-6. Typical Labor Rates per 100 Minutes.

Job Grade	Rate
A	$3.94
B	3.76
C	3.58
D	3.41
E	3.18
F	3.15
G	3.04
H	$2.94

Factory Burden. A factory burden rate is used to cover such miscellaneous operations as handling, inspection, engineering time, machine setup time, production starting costs, standard cutting tools, etc., when these are not itemized in the estimate. Since nonproductive costs vary by department, the use of separate burden percentages is advisable, although some companies prefer to apply a standard overall figure. Using the latter approach, the sum of the direct labor costs for all machining and ancillary operations performed on the aluminum forging within the plant ($81.46) is multiplied by a standard burden rate of 210 per cent, resulting in a burden cost of $171.07.

Purchased Parts. The purchasing department furnishes information to the cost department showing that blank forgings will cost $1.85 each, and that the charge for anodizing by an outside vendor will be $65 per 100 pieces. As an expedient, these two items are grouped under "Materials" for a total of $250 per 100. Thus the material, labor, and burden (MLB) sum will be $502.53 per 100.

Selling Price

Next the estimate goes to the accounting department which establishes final selling price. Percentages are added for general and administrative costs (G and A) and for profit. These vary from one company to another, but the normal factors are 15 and 10 per cent, respectively.

General and Administrative Costs. First, the G and A allowance is applied by multiplying the MLB sum by this company's figure of 15 per cent, resulting in a G and A cost of $75.38. Adding this to the MLB amount gives a total cost of $577.91 per 100 pieces.

Profit

To apply the profit percentage (10 per cent) to derive the selling price, this formula is used:

$$\text{Selling Price} = \frac{\text{Total Cost}}{90\% \ (100\% - 10\% \ \text{profit})}$$

The resulting selling price for the aluminum housing is $642.12 per 100 pieces, or $6.42 per piece.

Customer Quotation

After the selling price and applicable discount allowances are entered on the action copy of the estimate, the action copy goes to the sales department, and sales prepares a formal proposal for the customer. The proposal indicates the company's intent to meet requested quantity and delivery specifications, or may suggest alternatives deemed necessary, or to the customer's advantage. It also shows the price of the article (usually per 100 pieces) with applicable discounts, the cost of tooling if quoted separately, and a payment schedule. Prices are usually quoted f.o.b. the manufacturer's address but, if not, cost and type of transportation are also specified.

Profit. To arrive at selling price, the total price per hundred ($577.91) is multiplied by the appropriate profit allowance (10 per cent), and the resulting figure is added to total cost. The resulting selling price for the aluminum housing is $642.12 per hundred, or $6.42 each.

SCREW MACHINE COST ESTIMATING

Parts made on automatic screw machines require complicated calculations to compute machine time. This example explains the steps necessary to derive such times and determine the other necessary manufacturing costs for the two parts shown in Fig. 5-7.

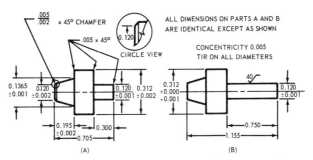

Fig. 5-7. Two pins made on an automatic screw machine.

Part Analysis

At first glance, the two parts (Fig. 5-7) seem identical except for shank length. However, closer analysis shows that the short pin, View *a*, has a tolerance of ±.002 in. on the .312-in. diameter, whereas the long pin, View *b*, has a tolerance of $^{+.000}_{-.001}$ in. on the same diameter.

Raw Stock. The ±.002-in. tolerance on the short pin will permit the use of mill run cold-drawn steel. To maintain the $^{+.000}_{-.001}$-in. tolerance on the longer pin, it will be necessary to use $^{21}/_{64}$-in.-diameter stock and box-turn or cross-slide form the .312-in. diameter. Larger-diameter stock should also be used on parts with high concentricity and straightness requirements. However, the larger-diameter stock may necessitate using a larger machine.

Processing Requirements. Further study of the part in View *b* indicates a slenderness or shank length-to-diameter ratio of 6.25:1. When forming a diameter using cross-slide tools, this ratio should not normally exceed 3:1; thus the shank on the part of Fig. 5-7*b* must be turned with a turret tool instead of forming with a cross-slide tool. To produce the required tolerance of ±.001 in. and the surface finish of 40 micron., a rough and finish cut will be required.

Short Pin

The manufacturing costs associated with producing the short pin (shown in Fig. 5-7*a*) are estimated and summarized on the "Machining Estimate" form (Fig. 5-8). The top portion of this form is used to develop labor and burden costs, and the bottom portion to add material, tooling, and general and administrative costs.

Tool Layout. The complete list of tools required for this part and their hypothetical costs are:

$^{5}/_{16}$-in.-diameter collet	$ 15.00
$^{5}/_{16}$-in.-diameter feed finger	8.00
Stock stop for turret	3.50
Front cross-slide circular form tool holder	15.00
Rear cross-slide circular form tool holder	15.00
Front cross-slide circular form tool	20.00
Rear cross-slide circular form tool	16.00
Set of cams consisting of front cross-slide cam, rear cross-slide cam and turret slide cam	60.00
	$152.50

MACHINING ESTIMATE

		INITIAL	DATE
COMPANY NAME *Electronics American Corp.*		INQUIRY BY *G.B.C.*	*12-5-68*
ADDRESS *12345 B. Street*		ESTIMATE BY *J.J.J.*	*12-6-68*
Midvale, Pa.		CHECKED BY *J.J.J.*	*12-6-68*
		APPROVED BY *J.J.*	*12-7-68*

PROJECT NO. *5401*

OP NO	OPERATION	MACHINE	NO. IN CREW	PROD'N PCS/HR	TIME HRS/SU	TIME SET UP MIN/PC	TIME MINUTES PER PC	TOTAL MIN/PC	LABOR AVG RATE $1 MIN	LABOR COST PER PC	BURDEN RATE %	BURDEN PER PC	TOTAL COST
10	*Form, form & Cutoff*	*#00 H.S. B.& S. Auto Scr. Mach.*	1	900	2.0	0.002	0.0196	0.0216	$0.0022	0.0010	225	0.0068	0.0078

This is for one man operating 4 machines

Full time is 0.0667 min.

Burden applied to full time

MATERIAL:

Base 100# $7.65	
Chem.	*1.35*
Size	*2.90*
Quan.	*2.00*
Total	*$13.90*

Total Labor + burden	*0.0078*
Material Cost	*0.0025*
Manufacturing Cost	*0.0103*
O + A (40% of Mfg. cost)	*0.0041*
Total cost	*0.0144*
Profit (15%)	*0.0022*
Selling Price, each $0.0166	
100 pcs. $1.66	
Special tool cost $96.00	

Fig. 5–8. Machining estimate for short pin (Fig. 5–7a).

This is a complete tool estimate; however, all the items except the circular form tools and cams are usually purchased with the machine, and only the cost of these latter tools ($96) is included in the estimate (see Fig. 5–8).

A sketch of the short pin and an outline of the tools used to machine it are shown in Fig. 5–9. The front slide Tool *A* forms the tapered portion, the chamfer on the left-hand side of the large diameter, and necks the cutoff area

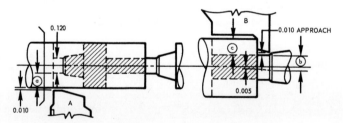

Fig. 5–9. Tool layout for workpiece (Fig. 5–7a).

to a .120-in. diameter. The rear slide Tool *B* forms the .120-in.-diameter shank and cuts off the previously formed part.

Machine Time. Determining the machine time necessary to fabricate this part requires the following steps:

1) Determine the proper spindle speed to machine the specified material.
2) Figure the throws of the cam lobes and the spindle revolutions required for the cutting operations.
3) Determine which operations can be overlapped.
4) Figure spindle revolutions required for the nonproductive operations such as feeding of stock and indexing of the turret.

5) Determine operation sequence by the following procedure: (a) provide clearance for cross-slide and turret tools, (b) find total estimated spindle revolutions required to finish the piece and select actual revolutions available with regular change gears that are nearest to the estimated number, and (c) adjust estimated spindle revolutions for each operation so that the total will equal the actual number available on the machine.

Speeds and Feeds. The recommended speeds for machining steels are given in Table V-7. From this table the cutting speed for AISI C1213 or B1113 steels is found to be 225 sfpm. From a table of standard data listing rpm.'s for cutting speeds and diameters, it can be interpolated that a $5/16$-in.-diameter workpiece rotating at 225 sfpm. has an approximate rpm. of 2,750. Referring to the production table for the No. 00 automatic screw machine (Table V-8), the selected speed would be 3,085 rpm. as the nearest speed available to the suggested speed of 2,750 rpm.

The suggested speeds for standard tools on single-spindle screw machines are given in Table V-9. Under Column E, a form tool $1/4$ to $3/8$ in. wide has a suggested feed of .0009 ipr. A circular cutoff tool has a feed of .0013 ipr. Since a forming operation is performed with the cutoff operation, the forming tool feed of .0009 ipr. will apply.

Spindle Revolutions. The number of spindle revolutions for each cutting tool is estimated as follows:

For Tool *A* the dimension *a* must be calculated.

Stock diameter	.312 in.
Formed diameter	−.120 in.
Difference is 2*a*	.192 in.
Dimension *a* is	−.096 in.
	.096 in.
Approach	+.010 in.
Total tool travel	.106 in.

Spindle revolutions for Tool *A* is .106 in. ÷ .0009 ipr. = 117.7 or 118 revolutions.

For Tool *B*, dimensions *b* and *c* must be calculated and the larger used to determine the number of spindle revolutions.

Dimension *b*		Dimension *c*	
Formed diameter	.120 in.	Stock diameter	.312 in.
$1/2$ formed diameter	.060 in.	Formed diameter	+.120 in.
Dimension D from Table V-10	.011 in.	Difference is 2*c*	.192 in.
Tool travel past center	.005 in.	Dimension *c*	.096 in.
Approach	+.010 in.	Approach	+.010 in.
Total tool travel	.086 in.	Total tool travel	.106 in.

The number of spindle revolutions for Tool *B* is .106 in. ÷ .0009 ipr. = 117.7 or 118.

Overlapping Operations. The cutting operations should be performed simultaneously for the greatest production economy. Since Tool *A* is cutting closer to the spindle than Tool *B*, Tool *A* should not generate a diameter (in this case,

Table V-7. Cutting Speed and Machinability Rating for Steels.

AISI No.	SAE No.	Speed (sfpm.)	Relative Speed (per cent)*	AISI No.	SAE No.	Speed (sfpm.)	Relative Speed (per cent)*
C1145 annealed	1145	130	78	C1050	1050	90	54
C1146	1146	115	70	C1050 annealed	1050	115	70
C1151	1151	115	70	C1051	....	90	54
C1151 annealed	1151	135	81	C1052	1052	80	49
C1211	....	155	94	C1053	....	90	54
C1212	....	165	100	C1054	....	90	54
C1213 (RSC 1213)	....	225	136	C1055 annealed	1055	85	51
				C1059 annealed	....	85	51
Carbon Steels				C1060 annealed	1060	85	51
				C1061 annealed	....	85	51
				C1064 annealed	....	80	49
C1008	1008	110	66	C1065 annealed	1065	80	49
C1010 (light feeds)	1010	120	....	C1066 annealed	....	80	49
C1011 (light feeds)	....	120	....	C1069 annealed	....	80	49
C1012 (light feeds)	....	120	....	C1070 annealed	1070	80	49
C1013 (light feeds)	....	120	....	C1071 annealed	....	80	49
C1015	1015	120	72	C1074 annealed	....	75	45
C1016	1016	130	78	C1075 annealed	....	75	45
C1017	1017	120	72	C1078 annealed	1078	75	45
C1018	1018	130	78	C1080 annealed	1080	70	42
C1019	1019	130	78	C1084 annealed	1084	70	42
C1020	1020	120	72	C1085 annealed	1085	70	42
C1021	1021	130	78	C1086 annealed	1086	70	42
C1022	1022	130	78	C1090 annealed	1090	70	42
C1023	....	125	76	C1095 annealed	1095	70	42
C1024	....	110	66	B1010	....	130	78
C1025	1025	120	72				
C1026	1026	130	78	*Alloy Steels*			
C1027	1027	110	66				
C1029	....	115	70				
C1030	1030	115	70	1330 annealed	1330	100	60
C1031	....	115	70	1335 annealed	1335	100	60
C1033	1033	115	70	1340 annealed	1340	95	57
C1034	....	115	70	3140	3140	70	42
C1035	1035	115	70	3140 annealed	3140	110	66
C1036	1036	105	64	E3310 annealed	3310	85	51
C1037	1037	115	70	4012	4012	130	78
C1038	1038	105	64	4023	4023	130	78
C1039	1039	105	64	4024	4024	130	78
C1040	1040	105	64	4027	4027	110	66
C1041	1041	95	57	4028	4028	120	72
C1042	1042	105	64	4037 annealed	4037	120	72
C1043	1043	95	57	4042 annealed	4042	115	70
C1045	1045	95	57	4047 annealed	4047	110	66
C1045 annealed	1045	120	72	4063 annealed	4063	85	51
C1046	1046	95	57	4118	4118	130	78
C1049	1049	90	54	4130 annealed	4130	120	72

Table V-7. Cutting Speed and Machinability Rating for Steels *(Continued).*

AISI No.	SAE No.	Speed (sfpm.)	Relative Speed (per cent)*	AISI No.	SAE No.	Speed (sfpm.)	Relative Speed (per cent)*
				Alloy Steels			
4135 annealed	....	115	70	8115	8115	115	70
4137 annealed	4137	115	70	8615	8615	115	70
4140 annealed	4140	110	66	8617	8617	110	66
4142 annealed	....	110	66	8620	8620	110	66
4145 annealed	4145	105	64	8622	8622	110	66
4147 annealed	....	105	64	8625	8625	105	64
4150 annealed	4150	100	60	8627	8627	105	64
4320 annealed	4320	100	60	8630 annealed	8630	120	72
4337 annealed	....	90	54	8637 annealed	8637	115	70
4340 annealed	4340	95	57	8640 annealed	8640	110	66
4422	4422	120	72	8642 annealed	8642	110	66
4427	4427	115	70	8645 annealed	8645	105	64
4520	4520	115	70	8650 annealed	8650	100	60
4615	4615	110	66	8655 annealed	8655	95	57
4617	....	110	66	8660 annealed	8660	90	54
4620	4620	110	66	8720	8720	110	66
4621	....	110	66	8735 annealed	8735	115	70
4718	4718	100	60	8740 annealed	8740	110	66
4720	4720	100	60	8742 annealed	....	110	66
4815 annealed	4815	85	51	8822	8822	105	64
4817 annealed	4817	80	49	9255 annealed	9255	90	54
4820 annealed	4820	80	49	9260 annealed	9260	85	51
5015	5015	130	78	9262 annealed	9262	80	49
5046 annealed	5046	115	70	E9310 annealed	9310	85	51
5115	5115	125	76	9840 annealed	9840	85	51
5120	5120	125	76	9850 annealed	9850	75	45
5130	5130	95	57	TS4150 annealed	....	100	60
5132 annealed	5132	120	72	TS14B35			
5135 annealed	5135	120	72	annealed	....	120	72
5140 annealed	5140	115	70	50B40 annealed	50B40	115	70
5145 annealed	5145	110	66	50B44 annealed	50B44	115	70
5147 annealed	5147	110	66	50B46 annealed	50B46	115	70
5150 annealed	5150	105	64	50B50 annealed	50B50	115	70
5155 annealed	5155	100	60	50B60 annealed	50B60	105	64
5160 annealed	5160	100	60	51B60 annealed	51B60	100	60
E50100 annealed	50100	70	42	81B45 annealed	81B45	110	66
E51100 annealed	51100	65	40	86B45 annealed	86B45	105	64
E52100 annealed	52100	65	40	94B15	94B15	115	70
6118	6118	110	66	94B17	94B17	110	66
6120	....	95	57	94B30 annealed	94B30	120	72
6150 annealed	6150	100	60	94B40 annealed	94B40	110	66

*Based on AISI B1112 as 100 per cent.

less than ¼ in.) before Tool *B* has finished forming the .120 ± .001 in. diameter. This ¼-in. dimension is based on the 3:1 slenderness ratio previously established.

Table V-8. Spindle Speeds and Revolutions per Piece for Laying Out Cams.
Production table, 240 rpm. driving shaft, No. 00 automatic screw machine

High spindle speeds in combination with low speeds

7200	5970	5035	4305	3715	3085	2580	2270	1915	1690	1425	1255	1050	870	750	640	540	450	Ratio
3130*	2600	2190	1875	1615		1125	990	835	735	620	545		380	325	280	235	195	2.3
2625*	2175	1835	1570	1355	1125		825	700	615	520		380	320	275	235	200	165	2.7
2305*	1910	1615	1380	1190	990	825		615	540		400	335	280	240	205	175	145	3
1945*	1615	1360	1165	1005	835	700	615			385	340	285	235	205	175	145	120	3.7
1715*	1425	1200	1025	885	735	615	540			340	300	250	210	180	155	130	105	4.2
1450	1200	1015	865	745	620	520		385	340		250	210	175	150	130	110	90	5
1270	1055	890	760	655	545		400	340	300	250		185	155	135	115	96	80	5.6
1065	885	745	635	550		385	335	285	250	210	185		130	110	95	80	67	6.7
885	735	620	530		380	320	280	235	210	175	155	130		92	79	67	55	8
765	635	535		395	325	275	240	205	180	150	135	110	92		68	58	48	9.3
650	540		390	335	280	235	205	175	155	130	·115	95	79	68		49	41	11
550		385	330	285	235	200	175	145	130	110	96	80	67	58	49		34	13
	380	320	275	235	195	165	145	120	105	90	80		67	55	48	41	34	16

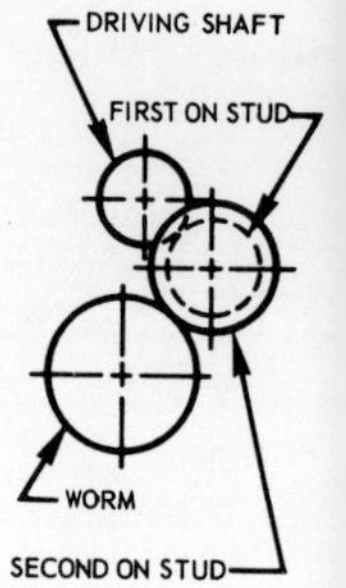

Revolutions at max speed to feed stock or index turret 1/4 sec.

30	25	21	18	15	13	11	9	8	7	6	5	4.4	3.6	3.1	2.7	2.3	1.9

Revolutions of spindle at max speed to make one piece

30	25	21	18	15	13	11	9	8	7	6	5	4.4	3.6	3.1	2.7	2.3	1.9	Time, sec.	Gross product per hr†	Gear on driving shaft	1st gear on stud	2d gear on stud	Gear on worm shaft	Hundredths of cam
90	75	63	54	46	39	32	28	24	21	18	16	13	11	9	8	7	6	¾	4800	100	20	80	60	34
105	87	73	63	54	45	38	33	28	25	21	18	15	13	11	9	8	6.5	⅞	4114	100	20	40	35	29
120	100	84	72	62	51	43	38	32	28	24	21	18	15	13	11	9	8	1	3600	80	20	50	40	25
150	124	105	90	77	64	54	47	40	35	30	26	22	18	16	13	11	9	1¼	2880	100	20	40	50	20
180	149	126	108	93	77	65	57	48	42	36	31	26	22	19	16	14	11	1½	2400	100	25	50	60	17
210	174	147	126	108	90	75	66	56	49	42	37	31	25	22	19	16	13	1¾	2057	80	35	50	40	15
240	199	168	144	124	103	86	76	64	56	48	42	35	29	25	21	18	15	2	1800	100	25	50	80	13
270	224	189	161	139	116	97	85	72	63	53	47	39	33	28	24	20	17	2¼	1600	80	20	25	45	12
300	249	210	179	155	129	108	95	80	70	59	52	44	36	31	27	23	19	2½	1440	80	20	25	50	10
330	274	231	197	170	141	118	104	88	77	65	58	48	40	34	29	25	21	2¾	1309	80	20	25	55	10
360	299	252	215	186	154	129	113	96	85	71	63	53	44	38	32	27	23	3	1200	80	20	25	60	9
390	323	273	233	201	167	140	123	104	92	77	68	57	47	41	35	29	24	3¼	1107	80	20	25	65	8
420	348	294	251	217	180	151	132	112	99	83	73	61	51	44	37	32	26	3½	1028	100	35	40	80	8
450	373	315	269	232	193	161	142	120	106	89	78	66	54	47	40	34	28	3¾	960	50	25	40	60	7
480	398	336	287	248	206	172	151	128	113	95	84	70	58	50	43	36	30	4	900	50	20	40	80	7
504	418	352	301	260	216	181	159	134	118	100	88	74	61	53	45	38	32	4⅕	857	100	35	25	60	6
540	448	378	323	279	231	194	170	144	127	107	94	79	65	56	48	41	34	4½	800	80	40	25	45	6
576	478	403	344	297	247	206	182	153	135	114	100	84	70	60	51	43	36	4⅘	750	100	40	25	60	5
600	498	420	359	310	257	215	189	160	141	119	105	88	73	63	53	45	38	5	720	80	40	25	50	5
660	547	462	395	341	283	237	208	176	155	131	115	96	80	69	59	50	41	5½	654	80	40	25	55	5
719	597	503	431	371	308	258	227	191	169	143	125	105	87	75	64	54	45	6	600	80	40	25	60	5
779	647	545	466	402	334	280	246	207	183	154	136	114	94	81	69	59	49	6½	553	80	40	25	65	4
839	697	587	502	433	360	301	265	223	197	166	146	123	102	88	75	63	53	7	514	50	35	40	80	4
899	746	629	538	464	386	323	284	239	211	178	157	131	109	94	80	68	56	7½	480	80	50	25	60	4
959	796	671	574	495	411	344	303	255	225	190	167	140	116	100	85	72	60	8	450	100	40	20	80	4
989	821	692	592	511	424	355	312	263	232	196	173	144	120	103	88	74	62	8¼	436	50	55	40	60	3
1079	896	755	646	557	463	387	340	287	254	214	188	158	131	113	96	81	68	9	400	50	45	40	80	3
1124	933	787	673	580	482	403	355	299	264	223	196	164	136	117	100	84	70	9¾	384	80	60	40	100	3
1199	995	839	718	619	514	430	378	319	282	238	209	175	145	125	107	90	75	10	360	80	40	25	100	3
1259	1045	881	753	650	540	452	397	335	296	249	220	184	152	131	112	95	79	10½	342	50	35	20	60	3

DRIVING SHAFT — FIRST ON STUD — WORM — SECOND ON STUD

*These combinations not to be used in opposite directions.
†Net will vary with factory conditions and the character of the work.

Table V-9. Approximate Cutting Speeds and Feeds for Standard Tools on a Single-Spindle Automatic Screw Machine.

Type of Tool	Cut Width or Depth (in.)	Cut Diameter of Hole (in.)	A	B	C	D	E	F	G	H	I
Boring Tools	.005	....	....	.0062	.0055	.005	.005	.0048	.0045	.0042	.004
Box Tools	1/32	....	.012	.0105	.0093	.0085	.0085	.0077	.0068	.0059	.005
One-chip finishing	1/16	....	.010	.0085	.0075	.0068	.0068	.0061	.0054	.0047	.004
V back rest – brass	1/8	....	.008	.0075	.0066	.006	.006	.0053	.0045	.0038	.003
Roller rest – steel	3/16	....	.008	.0063	.0056	.0051	.0051	.0044	.0036	.0028	.002
	1/4	....	.006	.0052	.0046	.0042	.0042	.0036	.0029	.0022	.0015
Finish cut – V rest	.005	....	.010	.010	.0093	.0085	.0085	.0079	.0073	.0066	.006
Center Drills	....	Under 1/8	.003	.0018	.0016	.0015	.0013	.0013	.0012	.0011	.001
		Over 1/8	.006	.0044	.0038	.0035	.003	.0031	.0027	.0024	.002
Cutoff Tools:											
Angular	....	....	.0015	.0008	.0007	.0006	.0005	.0005	.0005	.0004	.0004
Circular, over 1/8 diameter	3/64 to 1/8	....	.0035	.0018	.0016	.0015	.0013	.0013	.0012	.0011	.001
Straight, under 1/8 diameter	.020 to .040	....	.002	.001	.0009	.0008	.0007	.0007	.0006	.0006	.0005
Drills, Twist	....	.020	.0017	.0012	.0011	.001	.0009	.0009	.0008	.0007	.0006
		.040	.0024	.0017	.0015	.0014	.0012	.0012	.0011	.0009	.0008
		1/16	.0048	.0025	.0022	.002	.0017	.0018	.0016	.0014	.0012
		3/32	.0072	.0031	.0027	.0025	.0021	.0022	.002	.0018	.0016
		1/8	.011	.0044	.0038	.0035	.003	.0031	.0027	.0024	.002
		3/16	.014	.005	.0044	.004	.0034	.0038	.0035	.0032	.003
		1/4	.016	.0062	.0055	.005	.0043	.0045	.004	.0035	.003
		5/16	.016	.0069	.006	.0055	.0047	.005	.0045	.004	.0035
		3/8	.014	.0075	.0066	.006	.0051	.0055	.005	.0045	.004
		1/2	.012	.0075	.0066	.006	.0051	.0055	.005	.0045	.004
		5/8	.010	.0075	.0066	.006	.0051	.0055	.005	.0045	.004
Form Tool	1/8 and 1/4	....	.002	.0012	.0011	.001	.001	.0009	.0008	.0008	.0007
	3/8	....	.0018	.0011	.001	.0009	.0009	.0008	.0007	.0007	.0006
	1/2	....	.0015	.001	.0009	.0008	.0008	.0007	.0007	.0006	.0005
	5/8 and 3/4	....	.0012	.0009	.0008	.0007	.0007	.0006	.0006	.0005	.0004
	1 in. and up	....	.001	.0007	.0007	.0006	.0006	.0005	.0005	.0004	.0004
Balance Turning:											
Turned diameter under 5/32	1/32	....	.012	.012	.011	.010	.0085	.0095	.009	.0085	.008
	1/16	....	.010	.010	.0095	.009	.0077	.0082	.0075	.0066	.006
Turned diameter over 5/32	1/16	....	.012	.012	.0115	.011	.010	.0102	.0095	.0087	.008
	1/8	....	.012	.012	.011	.010	.0085	.0095	.009	.0085	.008
	3/16	....	.010	.010	.009	.008	.0068	.007	.0065	.006	
	1/4	....	.009	.0087	.0077	.007	.006	.0063	.0057	.0051	.0045
Knee Tool	1/64	....	.015	.015	.013	.012	.0105	.0115	.011	.0105	.010
	1/32	....	.012	.012	.011	.010	.0085	.0095	.009	.0085	.008
Knurl Tool	....	....	.020	.018	.0165	.015	.0127	.0137	.0125	.0122	.010
Turret on	....	....	.040	.037	.033	.030	.0255	.0287	.0275	.0262	.025
Turret off	....	....	.004	.0025	.0022	.002	.002	.002	.002	.002	.002
Side or swing	....	....	.006	.005	.0044	.004	.0034	.0037	.0035	.0032	.003
Top	....	....	.005	.0037	.0033	.003	.0026	.0027	.0025	.0022	.002
		....	.008	.0075	.0066	.006	.0051	.0055	.005	.0045	.004
Pointing and Facing	....	....	.001	.001	.0009	.0008	.0007	.0007	.0006	.0006	.0005
		....	.0025	.0025	.0022	.002	.0017	.0017	.0014	.0011	.0008
Reamers and Bits	....	Under 1/8	.007	.007	.0065	.006	.0051	.0055	.005	.0045	.004
	.003 to .004	....	.010	.010	.0088	.008	.0068	.0075	.007	.0065	.006
		Over 1/8	.010	.010	.010	.010	.0085	.009	.008	.007	.006
	.004 to .008	....	....	....	....	....	....	.0095	.009	.0085	.008
Recessing Tool, end cut	....	....	.001	.0008	.0007	.0006	.0005	.0005	.0005	.0004	.0004
		....	.005	.0037	.0033	.003	.0025	.0027	.0025	.0022	.002
Inside cut	1/16	....	.0025	.0025	.0022	.002	.0017	.0018	.0017	.0016	.0015
	1/8	....	.0008	.0008	.0007	.0006	.0005	.0005	.0005	.0004	.0004
Swing Tool, forming	1/8	....	.002	.0009	.0008	.0007	.0006	.0006	.0006	.0005	.0005
	1/4	....	.0012	.0006	.0006	.0005	.0004	.0004	.0004	.0003	.0003
	3/8	....	.001	.0005	.0004	.0004	.0003	.0003	.0003	.0002	.0002
	1/2	....	.0008	.0004	.0003	.0003	.0002	.0002	.0002	.0002	.0002
Turning Straight†	1/32	....	.008	.0075	.0066	.006	.0051	.0053	.0047	.0041	.0035
	1/16	....	.006	.005	.0044	.004	.0034	.0038	.0035	.0032	.003
	1/8	....	.005	.0037	.0033	.003	.0025	.0027	.0025	.0022	.002
	3/16	....	.004	.0031	.0027	.0025	.0021	.0022	.002	.0017	.0015

Note: Figures in this table are only approximate, to be used as a basis from which proper figures for the job in hand may be calculated. They are averages; if the work has any features out of the ordinary, take these into consideration and alter the figures accordingly. (*Footnotes continued on bottom of p. 72.*)

Table V-10. Angles and Thicknesses for Circular Cutoff Tools*.

A is 23 deg. when cutting brass, aluminum, copper, silver, and zinc.
A is 15 deg. when cutting steel, iron, bronze, and nickel.
Least thickness used when cutting off into tapped holes is the lead of two
and one-half threads plus .010 in.
Least thickness used when cutting off into reamed holes smaller than $\frac{1}{8}$
in. diam. is .040 in.
Thickness used when cutting off tubing is two-thirds *T* as given below for
corresponding diameters of stock.
Thickness used when angles or radii start from outside diameter of tool
is governed by varying conditions and determined accordingly.

Diameter of Stock	Thickness T	Depth of Angle D	
		For Brass	For Steel
$\frac{1}{16}$	.020	.0085	.0055
$\frac{3}{32}$	.030	.013	.008
$\frac{1}{8}$	.040	.017	.011
$\frac{3}{16}$	.050	.0215	.0135
$\frac{1}{4}$	.060	.0255	.016
$\frac{5}{16}$	.070	.030	.019
$\frac{3}{8}$	.080	.034	.021
$\frac{7}{16}$	.090	.038	.024
$\frac{1}{2}$ to $\frac{9}{16}$	.100	.042	.027
$\frac{5}{8}$ to $\frac{3}{4}$	.120	.051	.032
$\frac{13}{16}$ to 1	.140	.059	.038
$1\frac{1}{16}$ to $1\frac{5}{16}$	.160	.068	.043
$1\frac{3}{8}$ to $1\frac{7}{8}$	.190	.081	.051
2 to $2\frac{1}{2}$	.220	.093	.059

*All dimensions are in inches.

The tool travel for the overlapping operations is one-half the difference in diameters cut, plus the approach. Thus the tool overlap is:

$$\left(\frac{.312 - .250}{2} + .010\right) \div .009 = 45.5 \text{ or } 46 \text{ revolutions}$$

Revolutions for Nonproductive Operations. Since the part is machined completely from the cross slides, the only nonproductive operation requiring consideration is stock feeding. From past experience, stock feeding is estimated at 12 revolutions. A more accurate estimate will be made in the next step.

Operation Sequence. From the above calculations three operations for making this part have been established: feed stock, form with Tool *A*, and cut off with

---

* A. Free-cutting brass. Use maximum spindle speed available on machine.
  B. 2011-T aluminum, 800 spfm.
  C. 2017-T and 5052-T aluminum, 550 sfpm.
  D. 2024-T aluminum, 400 sfpm.; copper, 300 sfpm.; naval brass and Tobin bronze, 200 sfpm.
  E. High-sulfur steel B-1113, X-1112, and B & S 12A, 225 sfpm.; C-1113, 196 sfpm.; B-1112, 165 sfpm.
  F. Type 416 stainless steels and steels machined at 130 sfpm.
  G. Steels machined at 112 sfpm.
  H. Steels machined at 95 sfpm.
  J. All other stainless steels; tool steels, 75 sfpm.; monel, 60 sfpm.; phosphor bronze, high-speed steel, 65 sfpm.
  † Feeds for swing tools when turning a taper are the same as straight turning for the greatest depth of cut.

Tool *B*. These operations and number of spindle revolutions required are tabulated in Table V-11.

Table V-11. Machine Spindle Revolutions for Part in Fig. 5−7a.

Operation	Revolutions		
	Estimated	Adjusted	Actual
Feed Stock	12	14.4	16
Form (Tool *A*)	118	118.0	118
Form and Cut Off (Tool *B*) (118 − 46 overlap)	72	72.0	72
Total revolutions	202	204.4	206

The estimated number of spindle revolutions to complete the part is 202. In the production table for the No. 00 high-speed automatic screw machine (Table V-8) under the column headed 3085, the closest number is 206 revolutions; the same line of the right-hand column shows that .07 of the cam surface is required to feed the stock. For cam design calculations and estimating purposes, the 206 revolutions represent a complete cam surface having 100 divisions; or one division equals 2.06 revolutions. Multiplying .07 by 2.06 shows that 14.42 spindle revolutions are required to feed the stock. This is entered in the adjusted column of Table V-11 and gives a total of 204.4 spindle revolutions in the adjusted column, which is 1.6 less than the required 206 given in Table V-8. Since the stock must be fed in not less than 14.4 spindle revolutions, the 1.6 revolutions can be added to this item rather than to the machining operations.

From Table V-8, the time required to make one piece is 4 sec., or 900 pieces per hour at 100 per cent efficiency. This is equivalent to .0667 min. per piece.

Estimate Form. At this time, the estimator enters the accumulated data on the "Machining Estimate" (Fig. 5−8). This form provides space for a full description of each machining operation.

Labor. The labor cost per piece is determined by multiplying total time per piece by the labor rate ($.0422).

Total time per piece must be determined. On simple screw machine parts, operators will frequently tend three or four automatic screw machines. It is safe to estimate that an operator can tend at least two machines making more complex parts. In this example, one operator will be considered as tending four machines, with an allowance of 5 per cent for personal, fatigue, and tool trouble delays, or an efficiency of 85 per cent.

The number of minutes required per piece is .0667 ÷ .85 = .0785 min. Labor time required per piece per machine is .0785 ÷ 4 = .0196 min. This value is used as the direct labor time in the estimate. Setup time per piece of .002 is added to the direct labor time per piece (.0196) to give a total labor time per piece of .0216 min. Total labor time is multiplied by the labor standard ($.0422) to give a labor cost per piece of $.001.

Burden. Burden is determined by multiplying the burden rate (225 per cent) by the total machine time. Burden per piece is $.0068, and total labor and burden is $.0078. This cost is entered on the "Machining Estimate" (Fig. 5−8).

Material. The material costs for both the short and the long pin are based on the following prices:

Base per 100 lbs.	$ 7.65
Chemistry	1.35
Size	2.90
Quantity	2.00
Total for 100 lbs.	$13.90

Assuming 12-ft. bar stock and a 2-in. stub for the short pin, the number of pieces per bar is calculated as follows:

$$(144 - 2) \div (.750 + .040) = 179 \text{ pieces/bar}$$

$$144 \div 179 = .804 \text{ in. of stock per piece}$$

The weight of the $\frac{5}{16}$-in.-diameter steel is .022 lb/in. The rough weight is then .804 × .022 = .0177 lb/piece. Material cost is .0177 × $.139 = $.00246 per piece.

This material cost (rounded to $.0025) is added to total labor and burden on the "Machining Estimate" (Fig. 5–8) to arrive at the manufacturing cost for the part.

Overhead and Profit. General and administrative costs (overhead) and profit are taken as percentages of the manufacturing cost (excluding the "special tool" cost). General and administrative costs are obtained by multiplying the G and A factor (40 per cent) by the manufacturing cost of $.0103 per part. Next, G and A costs ($.0041) are added to the manufacturing cost ($.0103) to obtain total cost ($.0144). Profit is next computed by taking 15 per cent of total cost ($.0144) and adding the result to the total cost to obtain the selling price ($.0166).

When a quotation is made to the customer, the selling price is given as $1.66 per hundred, plus a special tool cost of $96.00.

Long Pin

Estimating machining time and the various costs associated with manufacturing the long pin shown in Fig. 5–7b requires the same steps described in the previous section. Although a "Machining Estimate" form is not included for this part, manufacturing data and costs should be recorded as in the previous section.

Tool Layout. A tool layout for this part is shown in Fig. 5–10. In turning the shank for this part, a rough cut to .140-in. diameter and a finish cut are taken

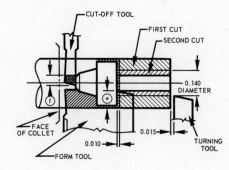

Fig. 5–10. Tool layout for workpiece of Fig. 5–7b.

with box tools mounted in the hexagon turret. Since the turning tool cannot satisfactorily finish to a square shoulder, .010 in. is left to be machined by a cross-slide form tool. An allowance of .015 in. is made for the approach of the turning tool. To maintain the required tolerance on the .312-in. diameter, $^{21}/_{64}$-in. diameter stock is used.

Machine Time. The length of cut for the turning tools is .750 + .015 − .010 = .755 in. The depth of cut is (.328 − .140) ÷ 2 = .094 in. From Table V-9, the rough turning feed is .0068 ipr.; for finish turning, the feed is .0085 ipr. The spindle revolutions required for the rough cut are .755 ÷ .0068 = 111 revolutions; for the finish cut, .755 ÷ .0085 = 89 revolutions. To obtain a favorable hill-to-valley relationship between the rough and finish cuts, the feed rate for finish turning was increased over that for rough turning. This was done because it would not be possible to increase the spindle speed satisfactorily for the finish turning operation.

Spindle Revolutions. Calculations for the front slide form tool and cutoff tool are:

Form Tool		Cutoff Tool	
Stock diameter	.328	Formed diameter	.120
Formed diameter	−.120	½ formed diameter	.060
Difference (2e)	.208	Dimension *D* from Table V-10	.011
Dimension *e*	.104	Tool travel past center	.005
Approach	+.010	Approach	.010
Total tool travel	.144 in.	Total tool travel	.086 in.

Spindle revolutions for form tool: .114 ÷ .0008 = 142.5 or 143

Spindle revolutions for cutoff tool: .086 ÷ .0013 = 66.1 or 66

Operation Sequence. The sequence of operations and number of spindle revolutions for each operation are shown in Table V-12.

Table V-12. Machine Spindle Revolutions for Part in Fig. 5−7b.

Operation	Revolutions		
	Estimated	Adjusted	Actual
Feed stock	15	15	15
Rough turn	111	111	110
Double index turret	23	24	24
Finish turn	89	89	88
Tool clearance for front slide	36	39	39
Form	143	143	141
Cutoff	66	66	65
Total revolutions	483	487	482

The four machining operations have a subtotal of 409 revolutions. From Table V-13, .025 (2½ hundredths) is required to index the turret. Using 4.5 revolutions per .01 of the cam surface, the number of revolutions to double

Table V-13. Hundredths Required to Index for Throws from Full-Height Cam.*
(Cut down plus cam throw)

4 to 5½ sec.		6 to 33 sec.		36 sec. and over	
Drop (in.)	Hundredths	Drop (in.)	Hundredths	Drop (in.)	Hundredths
11/32	2½	0 – 15/32	2½	0 – 5/8	2½
11/32– ½	3	15/32– 11/16	3	5/8 – 7/8	3
½ – 5/8	3½	11/16– 13/16	3½	7/8 –1 1/16	3½
5/8 – 3/4	4	13/16– 31/32	4	1 1/16–1 7/32	4
3/4 – 27/32	4½	31/32–1 3/32	4½	1 7/32–1 3/8	4½
27/32– 15/16	5	1 3/32–1 3/16	5	1 3/8 –1 ½	5
15/16 – 1/16	5½	1 3/16–1 5/16	5½		
1 1/16–1 1/8	6	1 5/16–1 3/8	6		
1 1/8 –1 1/4	6½	1 3/8 –1 7/16	6½		

*5½-in. cam on No. 00 machine.

index the turret is $2 \times 2.5 \times 4.5 = 22.5$ or 23 revolutions. The required clearance, for turret and cross-slide tools, is .08 (see Table V-14). The number of revolutions is $8 \times 4.5 = 36$. The subtotal is now 468 revolutions. From Table V-8 under the 3085 rpm. column, across from 463, .03 is required to feed the stock, or about 15 revolutions. The total estimated revolutions is 483. Referring to Table V-8, the nearest number of revolutions is 482. The estimated revolutions must now be adjusted to this new value:

Feed stock:	$4.82 \times 3 = 14.46$ or 15 revolutions
Double index turret:	$2 \times 2.5 \times 4.82 = 24.1$ or 24
Tool clearance:	$8 \times 4.82 = 38.56$ or 39

The total of the adjusted column is 487 or 5 revolutions above the required 482. The revolutions for the nonproductive operations are fixed; therefore the 5 revolutions must be deducted from the machining operations as shown in the right-hand column of Table V-12. The time required to make the piece is 9⅜ sec., or 384 per hour at 100 per cent efficiency. At 85 per cent efficiency, the time is .1254 min. The direct labor time when one man is operating 4 machines is .0314 min.

Labor and Burden. The direct labor time (.0314 min.) is multiplied by the labor rate to arrive at labor cost per piece, and the burden rate is applied to the resulting cost figure to give a total labor and burden cost.

Material. The cost per cwt. of bar stock is the same as for the short pin, $13.90. Material cost is calculated as follows:

$$(144 - 2) \div (1.155 + .040) = 110 \text{ pieces/bar}$$

$$144 \div 110 = 1.309 \text{ in. of stock per piece}$$

The weight of the 21/64-in.-diameter steel is .024 lb/in. The rough weight is $1.309 \times .024 = .0314$ lb/piece. The material cost is $.0344 \times .139 = \$.0044$ per piece.

Table V-14. Clearance in Hundredths between Turret Tools and Cross-Slide Circular Tools (2).

Tool No.	Turret Tools	No. 00	No. 0	No. 2	4	6	No. 00	No. 0	No. 2	4	6
		Front Cross-Slide Tool					*Back Cross-Slide Tool*				
BA-00C	Balance turning tool	6					6				
BA	Balance turning tool		6					6			
C-20D-22B-22G	Balance turning tool		7	7				7	7		
A-24L-26	Balance turning tool				6	6				6	6
K-20K-22D	Box tool	8	7	7			6	7	6		
A-24L-2G	Box tool				7	7				6	6
L-20L-22G	Box tool	6	6	6			7	7	7		
BM-20BA-22BA	Box tool	8	7	7			5	5	4		
CA-20CM-22AA	Box tool	8	7	7			6	6	6		
DA-20H-22G	Box tool	5	7	7			7	6	6		
EB-00FB	Box tool	8					6				
EB-20FB	Box tool		7				6				
EB-22FB	Box tool			7					6		
D-00CA	Centering and face tool	8					4				
D-11BA	Centering and face tool		7				3				
D-22CA	Centering and face tool			7					3		
	Self-opening die head	8	7	7	7	7	8	7	7	7	7
AB-20AB-22AB	Combination right- and left-hand knee tool	6	5	5			4	4	4		
D-20D-22DA	Knurl holder	7	6	6			7	6	6		
A-26	Knurl holder				6	6				5	5
BA-20C-20DA-22BA-22DA	Pointing tool	7	7	7			5	6	6		
CA	Pointing tool	8					6				
DA	Pointing tool	5					7				
A-24L-26	Pointing tool				6	6				6	6
BA-26B	Turret tool post				3	3					

NOTE: For No. 00 size machines, double the time given in table for No. 00 size machine, if a 3-sec. job or faster. On a 4-sec. job add hundredths, and on a 5-sec. job add 4 hundredths of cam surface to figures given in table for these machines.
*Not given.

ESTIMATING SAND CASTING COSTS

To estimate castings, the estimator should acquaint himself with material specifications, heat treatment specifications, final inspection requirements, and the casting design as specified in the cost estimate request. Material and heat treatment specifications, which add to the cost of castings, are often called out in notes on engineering drawings without being mentioned in the estimate request. Some companies establish their own code numbers for material specifications and for subsequent heat treatment and inspection operations with which the estimator should be familiar, as well as understanding the chemical composition and physical specifications of the materials to be used. He should also check customer specifications against foundry capabilities to ensure that acceptable finished castings can be produced.

The total cost of making castings is comprised of the following cost items: (1) material, (2) foundry tooling, (3) molding costs, (4) core costs, (5) machining and cleaning costs, (6) heat treatment costs, (7) inspection costs, and (8) foundry burden.

Material Costs

To determine the material cost of finished castings, the estimator calculates finished casting weight and multiplies this weight by the cost per pound of the metal used in the finished castings.

Finished Casting Weight. The estimated finished casting weight (the weight of the casting as shipped to the customer) is computed from engineering drawings and is simplified when the drawings indicate stock allowances if any surfaces are to be machined. When no such drawings are available, they should be prepared by the estimator and submitted to the customer or in-house requesting department for approval before proceeding with the estimate.

Finished casting weight is calculated by multiplying the volume of the casting (in cubic inches) by the weight per cubic inch of the material. While the volume of regularly shaped castings can usually be determined by the use of handbook formulas for geometrical solids, many cast parts are irregularly shaped, requiring complex calculations. A part of irregular shape and thickness should be divided into simple geometric segments, the volume of each determined as above by handbook formulas, and the total volume derived by adding the volume of each segment. For castings having irregular cross-sectioned areas, a planimeter (a precision instrument designed to assist in calculating the area of a plane surface by tracing its perimeter) can be used.

Cost of Metal in Finished Castings. Determining the cost of the metal used in finished castings involves the following steps:

1) Calculate the amount of metal charged into the furnace.
2) Make allowance for metal lossage.
3) Determine the amount of metal returned to the furnace for remelting.
4) Determine the cost/lb of metal poured from the furnace.

Furnace Charge. A shop yield factor based on previous foundry experience is used to compute the required foundry charge. Shop yield is the ratio of finished casting weight to the weight of the metal charged into the furnace. For example:

$$\frac{545,000 \text{ lbs. finished castings}}{1,000,000 \text{ lbs. metal charged}} = 54.5\%$$

Using a shop yield of 54.5 per cent, and assuming a finished casting weight of 10 lbs., the furnace charge per casting is computed as follows:

$$\frac{10 \text{ lbs.}}{54.5\% \ (.545)} = 18.35 \text{ lbs. charged metal}$$

Because shop yield varies according to the kind of casting produced, the estimator is well advised to keep a record of casting weights such as is shown in Tables V-15 and V-16.

Metal Lossage. In all foundries, a certain amount of the metal charged into the furnace is lost due to oxidation, spills, overruns, and gate cutoff. A typical allowance is 10 per cent of finished castings, or 5.4 per cent using a shop yield of 54.5 per cent.

For a casting weighing 10 lbs., the weight of lost metal is determined as follows:

$$10 \text{ lbs.} \times 10\% = 1 \text{ lb. of metal lost}$$

Table V-15. Record of Weights of Gray Iron Castings.

Part	1 Pieces per Mold	2 Shipping Weight (lbs/mold) (= 4 − 3)	3 Grind Loss (lbs.)	4 Yield Weight (lbs.) (= 7 − 6)	5 Yield Weight (per cent) (= 4 ÷ 7)	6 Gate and Sprue (lbs.)	7 Pouring Weight (lbs.)	8 Rough Casting Weight (lbs.) (= 4 ÷ 1)	9 Finish Casting Weight (lbs.) (= 2 ÷ 1)
Casting	1	33.125	.104	33.229	81.2	7.709	40.938	33.229	33.125
Transmission case	1	39.156	.157	39.313	74.9	13.187	52.500	39.313	39.156
Transmission case	1	51.750	.250	52.000	78.5	14.250	66.250	52.000	51.750
Pump body	2	13.604	.334	13.938	65.2	7.437	21.375	6.969	6.802
Pump body	4	28.332	....	28.332	76.1	8.918	37.250	7.083	7.083
Carrier	1	33.896	....	33.896	67.1	16.612	50.508	33.896	33.896
Bearing cap	2	4.084	....	4.084	67.1	2.002	6.086	2.042	2.042
End bell	4	13.124	....	13.124	58.3	9.376	22.500	3.281	3.281
End bell	2	6.750	....	6.750	53.9	5.764	12.514	3.375	3.375
Housing	4	14.812	....	14.812	71.8	5.813	20.625	3.703	3.703
Distributor base	4	15.452	....	15.452	70.4	6.486	21.938	3.863	3.863
Brake drum	1	78.000	....	78.000	85.2	13.500	91.500	78.000	78.000
Flywheel	1	34.000	....	34.000	79.5	8.750	42.750	34.000	34.000
Pump rotor	8	5.248	.128	5.376	45.3	6.499	11.875	.672	.656
Oil pump	6	13.500	.186	13.686	65.2	7.314	21.000	2.281	2.250
Cover	10	12.100	....	12.100	63.7	6.900	19.000	1.210	1.210
Pressure plate	1	11.125	....	11.125	70.9	4.563	15.688	11.125	11.125
Manifold	3	30.000	....	30.000	71.0	12.250	42.250	10.000	10.000
Block	1	190.584	1.916	192.500	74.9	64.500	257.000	192.500	190.584
Drum	1	80.719	....	80.719	85.6	13.531	94.25	80.719	80.719
Block	2	392.334	.486	392.820	84.0	74.667	467.450	196.410	196.167
Housing	1	37.125	.375	37.500	77.3	11.000	48.500	37.500	37.125
Outlet	8	10.440	....	10.440	51.6	9.810	20.250	1.305	1.305
Weight	2	14.750	....	14.750	65.7	7.688	22.438	7.375	7.375

Table V-16. Record of Weights of Malleable Iron Castings.

Part	Pieces per Mold	Shipping Weight (lbs/mold)	Grind Loss (lbs.)	Yield Weight (lbs.)	Yield Weight (per cent)	Gate and Sprue (lbs.)	Pouring Weight (lbs.)	Rough Casting Weight (lbs.)	Finish Casting Weight (lbs.)
Spacer	8	12.304	.152	12.456	79.6	3.200	15.656	1.557	1.538
Spacer	8	14.064	.184	14.248	79.2	3.752	18.000	1.781	1.758
Yoke	2	2.875	.031	2.90	43.0	3.852	6.758	2.906	2.875
Link	6	11.436	.690	12.126	56.4	9.374	21.500	2.021	1.906
Hinge	3	8.064	.063	8.127	66.7	4.061	12.188	2.709	2.688
Link	2	6.094	.062	6.156	35.7	11.094	17.250	3.078	3.047
Bearing sleeve	2	19.750	....	19.750	63.5	11.329	31.079	9.875	9.875
Spool	4	38.416	....	38.416	48.9	40.209	78.625	9.604	9.604
Spool	4	20.252	....	20.252	55.5	16.248	36.500	5.063	5.063
Washer	6	21.702	....	21.702	75.5	7.048	28.750	3.617	3.617
Wedge	10	5.030	.130	5.160	59.0	3.590	8.750	.516	.503
Casting	4	18.064	.188	18.252	80.2	4.498	22.750	4.563	4.516
Casting	1	17.834	.166	18.000	53.3	15.750	33.750	18.000	17.834
Rail block	1	30.375	.250	30.625	57.8	22.375	53.000	30.625	30.375
Rail brace	1	72.500	.500	73.000	82.0	16.000	89.000	73.000	72.500
Cap	6	16.782	.186	16.968	56.3	13.157	30.125	2.828	2.797
Bearing cap	1	27.875	.125	28.000	45.2	34.000	62.000	28.000	27.875
Clamp	5	4.570	.120	4.690	41.7	6.560	11.250	.938	.914
Clamp	12	3.132	.096	3.228	40.0	4.835	8.063	.269	.261
Gross head	1	28.800	.450	29.250	53.7	25.250	54.500	29.250	28.800
Cap	1	20.500	.094	20.594	65.9	10.656	31.250	20.594	20.500
Bearing cap	1	27.875	.125	28.000	45.2	34.000	62.000	28.000	27.875
Differential case	5	98.750	.690	99.440	63.5	57.060	156.500	19.888	19.750

Remelted Metal. Any metal not lost or consumed in the finished casting is returned to the furnace for remelting. The percentage of the metal charged into the furnace that is returned for remelting is then determined as follows:

Remelted Metal = Metal Charged − (Shop Yield + Metal Lost)
Remelted Metal = 100% − (54.5% + 5.4%)
          = 40%

For a finished casting weighing 10 lbs., and requiring a metal charge of 18.35 lbs., the amount of remelted metal would be calculated as follows:

Remelted Metal = Furnace Charge per Casting × Remelt Factor
Remelted Metal = 18.35 × .40
          = 7.34 lbs.

Cost of Poured Metal. The cost of the metal poured from the furnace is the sum of: (1) the cost of the metal charged, and (2) the cost of labor and overhead to charge the furnace and melt the metal.

The cost of the charged metal is based upon the current price of the metal or metals being used, plus the value of remelted metal.

The furnace labor and overhead charges are usually based upon costs derived from a previous accounting period. For example, the cost of labor and overhead for charging the furnace and melting the 1,000,000 lbs. of metal is $10,000. Then, $10,000 ÷ 1,000,000 lbs. = $.01/lb.

The $.01 value may require adjustment if either the anticipated labor and overhead costs for the current accounting period or the amount of metal to be poured is expected to change.

The cost of the poured metal per casting is determined as follows:

Poured Metal Cost per Casting = (Furnace Labor and Overhead + Cost of Metal Charged) × Casting Poured Weight

Where: Casting poured weight (10-lb. finished casting) = $18.35 lbs.
      Cost of metal charged   = $ .06/lb
      Furnace labor and overhead   = $ .01/lb

Poured Metal Cost per Casting = ($.01 + $.06) × 18.35 lbs. = $1.28

Finished Casting Metal Cost. Next the estimator determines the appropriate cost to assign the material actually used in the finished castings.

Cost of Metal in Finished Casting = Poured Metal Cost per Casting − (Amount of Remelted Metal × Value of Remelted Metal)

Where: Weight of finished casting   = 10 lbs.
      Poured metal cost per casting = $1.28
      Amount of remelted metal   = $7.34 lbs.
      Value of remelted metal   = $ .04 lb.

Cost of Metal in Finished Casting = $1.28 − (7.34 lbs. × $.04) = $.99

To obtain the cost per pound of metal used in the finished casting, the $.99 value is divided by the finished casting weight of 10 lbs. The resulting material cost per pound is $.099.

Foundry Tooling

Foundry tooling includes patterns, pattern plates, blow plates, and flasks, as well as various types of core-making tooling. The cost of foundry tooling required for a particular job is estimated and the cost added to the overall estimate. A tooling cost per casting is obtained by dividing total tooling cost by the number of castings produced.

Core-Making Tooling. Core-making tooling is generally estimated separately, and the cost of these tools is used to develop a separate core estimate. The core estimate is then added to the overall casting estimate.

Patterns. A pattern is set in the molding sand and sand is packed around it to produce the impression into which hot metal is poured to produce castings of a desired shape. The cost of patterns is assigned directly to the part being cast because patterns designed for one job are not generally usable for future orders.

Pattern Plates. This is the plate that separates the two halves of a pattern during molding. Composite pattern plates reduce costs by permitting the use of two or more different patterns where production quantities are low relative to the foundry production rate.

Patterns are easily removed from the composite pattern plate and can be replaced in any combination that will permit meeting the schedule for fulfilling casting orders. With such an arrangement, pattern costs may be kept at a minimum when volumes are low enough to permit making a small amount of patterns and mounting them with other patterns on existing pattern plates.

The cost of composite pattern plates should be split among the various casting orders produced with them.

Flasks. Flasks, the containers for the molding sand, are made in many different styles, types, and sizes. The size of the flask, its method of construction, and the material used in constructing it, all affect costs. When an estimate is being developed, the estimator must decide which flask is to be used. The costs of flasks designed and constructed especially for a particular job must be assigned to the castings produced for that job. The cost of permanent flasks, however, is generally included in foundry burden.

Molding Costs

Molding costs include the cost of preparing sand molds and the costs of pouring the hot metal into these molds. Molding cost is generally expressed as a cost per pound of acceptable finished castings. For example, a foundry produced 545,000 lbs. of acceptable finished castings with molding costs of $27,250 during one accounting period. The cost per pound of castings would be $.05. This cost would be used during the next accounting period to estimate various production lots of castings.

The number of sand molds that can be made and poured per hour is affected by the flask size, casting weight, type of pattern equipment, and number of cores to be set. A more accurate estimate can be derived by determining how many molds can be made per hour and then converting this value into the cost per pound of casting. The number of molds per hour can be established by using standard time data or time required on similar molds.

Core Costs

A core is a shaped projection of sand or other material inserted into the mold to create a cavity or recess in the casting. Dry-sand cores are formed separately and inserted after the pattern is removed but before the mold is closed.

Core Tooling and Equipment. A fairly accurate plan or layout of the cores in a particular casting should be used to estimate the cost of:

1) The core box used to make the cores
2) The driers necessary to support the cores during baking
3) The blow plate which may be required for a particular core-blowing machine
4) Racks, boxes, or special containers in which to store finished cores until used in the molding process
5) Fixtures required for the more complicated core assemblies
6) Fixtures, tanks, pumps, or filters that may be required for dipping or spraying special core washes
7) Core-pasting fixtures
8) Ovens to bake cores or dry-pasted assemblies.

Estimating Methods. The cost of cores can be determined by one of three different methods, depending upon the core size and method of making the core.

One method relies upon direct labor cost per core. The number of acceptable cores made during an accounting period is divided by the cost of making the cores which is direct labor cost only. The cost per core is then adjusted by adding overhead, which includes the cost of sand, core baking, supplies, supervision, etc.

The second method is the same as the first except that sand cost is excluded from overhead and assigned to the cores on an individual basis. This method reduces the overhead percentage and takes into account the cost of material. Generally, it results in a greater cost per core for large cores and a smaller cost per core for small cores than the first method.

A third method is to determine the direct labor cost per core according to the method by which it is made, e.g., bench, blower, etc., and then add overhead and sand cost.

Cleaning and Machining Costs

Rough sand castings are generally cleaned before shipping even if no machining is required. Many casting orders call for rough or finish machining to specifications.

Cleaning. The method used to estimate cleaning costs depends upon the type of cleaning operation. Hand or table blasting costs can be estimated by using standard time data. For tumbling, it is usually necessary to estimate the cost on a per-pound basis. If the sizes and weights of the castings vary considerably, the castings may be classified into groups and a cost factor determined for each classification.

Machining. The application of direct labor costs and overhead is usually more practical than using a cost-per-pound basis. Standard time data can easily be determined for machining operations such as grinding, chipping, filing, etc., and the direct labor hours estimated from these data.

Heat Treatment

For iron castings, heat treating may be specified for stress relieving or to improve machinability, and iron alloy castings may be quenched or tempered to increase their hardness and wear resistance. Steel castings may be normalized, annealed, stress-relieved, quenched, and tempered.

Heat treatment costs may be estimated on the basis of: (1) casting weight and (2) treatment time.

Weight Basis. Under the casting weight method, the entire cost of operating the heat treatment department is considered in conjunction with the average amount of work performed during an accounting period, and a factor developed enabling this cost to be assigned on the basis of pounds of castings treated. For example, the entire cost of operating a small heat treatment facility is $63,000 per year, and 1,512,000 lbs. of castings are heat treated. The cost of operating this heat treating operation ($63,000) is divided by the weight of the castings treated (1,512,000 lbs.), resulting in a cost per lb. of $.04. This cost is used for estimating heat treating costs during the next accounting period.

Time Basis. Under the treatment time method, the total cost of operating the heat treatment facilities is divided by the total number of hours the facilities are operated. Assuming a total cost of $63,000 per year and a total annual operating time of 1,500 hours, the hourly cost of operating the facilities is $42. This hourly cost is then assigned to the total castings treated during a given time period. For example, a production lot of castings weighing 4,200 lbs. is heat treated for four hours. The heat treatment cost (hourly rate × hours of operations) is $168. The heat treatment cost is divided by the weight of the castings (4,200 lbs.), giving a heat treatment cost of $.04 per pound.

Inspection and Shipping Costs

The simplest method of estimating inspection and shipping costs is by the pound. In some cases, it is possible to apply direct labor costs to a job, but the operations may be so minor that this procedure is not practical. Any special handling or packaging required should be considered, however.

Foundry Burden

Foundry burden, applied on the basis of pounds of finished castings, is computed by the use of a burden factor supplied by the accounting department. For example, if 545,000 lbs. of finished castings are produced and the burden factor is $.11 per pound of finished castings, the burden cost for the production lot of castings is $59,950.

Sand Casting Estimate Example

The part shown in Fig. 5–11 is a water outlet for a gasoline engine. The material specified is a common grade of gray iron, and the specified annual production quantity is 5,000 pieces. Accompanying the cost request for this part was a cost request for a thermostat housing for the same engine. Because

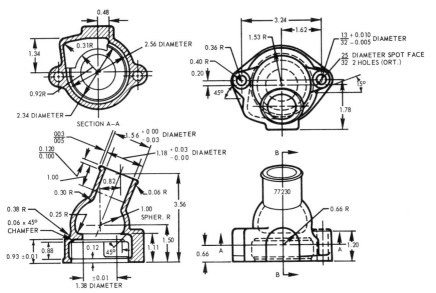

Fig. 5–11. Water outlet casting.

each engine assembly requires one of each type of casting, the process plan calls for making an equal number (four) of each casting at one time.

The mold was made on a pin-lift molding machine in a 13-in. by 18-in. flask, 5-in. drag and 6-in. cope, using a composite pattern with four impressions for the water outlet and four for the thermostat housing.

Material Cost. To determine material cost, the estimator multiplies finished casting weight by the cost per pound of the metal used in the finished casting.

Casting Weight. Casting weight is calculated from the engineering drawing (Fig. 5–11) by breaking the workpiece into suitable geometric sections and obtaining the volume of each. The total weight is found by multiplying the total volume by the weight of the material per unit volume.

Section 1 is the cylindrical top portion of the casting (Fig. 5–11): 1.56 in. outside diameter (O.D.), 1.18 in. inside diameter (I.D.), and 1.20 in. long. Section 2 is the spherical portion in the center of the casting. The third section is the base. The volume of the base is found by using a planimeter and Section A-A of Fig. 5–11. Table V-17 shows the calculation of the casting weight.

Table V-17. Calculation for Weight of Water Outlet Casting.

Material: Cast iron, .26 lb/cu in
Volume of section 1.
$$V = \pi h(r_1^2 - r_2^2)$$
From Fig. 5–11, $h = 1.2$ in.; $r_1 = .78$ in.; $r_2 = .59$ in.
$$V = 1.2\pi[(.78)^2 - (.59)^2] = 3.77(.61 - .35) = .98 \text{ cu. in.}$$
Volume of section 2.

$$V = \frac{4\pi}{3}(r_1^3 - r_2^3)$$

(Table continued next page).

Table V-17. Calculation for Weight of Water Outlet Casting (*Continued*).

From Fig. 5–11, $r_1 = 1.18$ in.; $r_2 = 1.00$ in.; $\dfrac{4\pi}{3} = 4.189$ in.

$V = 4.189[(1.18)^3 - (1)^3] = 4.189(1.64 - 1.00) = 2.68$ cu. in.
For spherical segment not required

$$V = \frac{\pi h^2}{3}(3r - h)$$

For upper portion:
When $r = 1.18$ in., $h = .3$ in.

$$V_1 = \frac{\pi(.3)^2}{3}[(3)(1.18) - .3] = .0945(3.24) = .306 \text{ cu. in.}$$

When $r = 1.00$ in., $h = .2$ in.

$$V_2 = \frac{\pi(.2)^2}{3}[3(1) - .2] = .042(2.8) = .118 \text{ cu. in.}$$

$V_1 - V_2 = .306 - .118 = .188$ cu. in.

For lower portion:
Assuming the spherical segment not wanted is .40 in. below the center line, when $r = 1.18$ in., $h = .78$

$$V_3 = \frac{\pi(.78)^2}{3}[(3)(1.18) - .78] = .64(2.76) = 1.765 \text{ cu. in.}$$

When $r = 1.00$ in., $h = .60$ in.

$$V_4 = \frac{\pi(.60)^2}{3}[3(1) - .60] = .415(2.4) = .905 \text{ cu. in.}$$

$V_3 - V_4 = 1.765 - .905 = .860$ cu. in.
Total volume of spherical portion:
  $2.68 - .860 - .188 = 1.628$ cu. in.
Volume of section 3 (using planimeter):
Area of large section $= 8.760$ sq. in.
Area of small section $= \underline{5.560}$ sq. in.
Area of metal $= 3.200$ sq. in.
$V = Ah = 3.2 \times 1.2 = 3.84$ cu. in.
Total volume of part $= .980 + 1.628 + 3.840 = 6.448$ cu. in.
Total weight of part $= 6.448 \times .26 = 1.686$ lbs.

Using the casting weight of 1.686 (as determined in Table V-17), and assuming a shop yield of 54.5 per cent, a remelt factor of 40 per cent, and a metal lossage factor of 10 per cent of finished casting weight, the following weights are calculated:

$$\text{Pouring Weight} = \frac{\text{Finished Casting Weight}}{\text{Shop Yield}}$$

$$= \frac{1.686 \text{ lbs.}}{54.5\%}$$

$$= 3.1 \text{ lbs.}$$

$$\text{Remelted Metal Weight} = \text{Pouring Weight} \times \text{Remelt Factor}$$

$$= 3.1 \text{ lbs.} \times 40\%$$

$$= 1.24 \text{ lbs.}$$

Lost Metal = Finished Casting Weight × Metal Lossage Factor
= 1.686 lbs. × 10%
= .169 lbs.

Finished Casting Metal Cost. The cost of the metal used in the finished casting is computed by the use of the following formulas:

(1) Poured Metal Cost per Casting = Pouring Weight per Casting × (Labor and Overhead + Charged Material Cost)
(2) Cost of Metal in Finished Casting = Poured Metal Cost per Casting − (Amount of Melted Metal × Values of Remelted Metal)

Where: Labor and overhead = $.01/lb
Charged material = $.06/lb
Remelted metal = $.04/lb

(1) Poured Metal Cost per Casting = 3.1 ($.01 + $.06) = $.217
(2) Cost of Metal in Finished Casting = $.217 − (1.24 lbs. × $.04) = $.167

Core Cost. The core-making operations are shown on the "Operation, Equipment, Tool & Gage, & Standard Time Routing" sheet in Fig. 5–12.

OPERATION, EQUIPMENT, TOOL & GAUGE, & STANDARD TIME ROUTING

LEGEND	ISSUE NO 1	SHEET 1 of 1	PART NAME Water Outlet Cores	PART NO 77230
	FORGING NO	ROUGH WEIGHT	MATERIAL	EFFECTIVE DATE 6/17/68
	CASTING NO	FINISH WEIGHT	REFERENCE	GROUP NO
BUDGET GROUP STD	CHANGE Orig.-Re-estimated due to diff. equip.		CURRENT GROUP STD	

TOOL & GAUGES	BUDGET STD 1961	REMARKS	OPER	OPER SEQUENCE & EQUIPMENT	STUDY	STD	NET HOURLY PROD	GROSS HOURLY PROD
	$.770	23 PW	5	HR A, B+C core & place	Est.	$1.54	Total/100 cores	
				on Conv. O.R. 1.385				
				Pc. Dump core box				
Note: This is a				Wires - O. Vents - 0			(90 boxes)	
Composite core box-			see	note / box / drier				
there is an A, B+C				3/plate cores/cstg. -				
core in a box, or				1 Oper - Man				
3 cores for 1 assy.								
GT 22907-285								
	.00364	23-02		All group operations		.00546	Total	
Core box WO 88470								
			8	Oven tenders & plate man		.00546		
			10	tin & pack		N.J. done w/opn.# 100		
		23 PW		Core & place on			Totals/100 cores	
				O.R.	100 cores			
				Pc. Core box				
				Wires - Vents -				
				1 box 1 drier / plate				
				Cores/cstg - Oper -				
	.00593	23-02		all group Operations		.0266	Total	
			100	Assemble "B" to "A" using		.0238	40.0	
				2 wires, paste "C" to "A" &	Est	.00281		
				"B" assy. & mud seam	11.8%			
BOOK NOS				CHANGED STANDARD			TOOLS	

Fig. 5–12. Operation, equipment, tool and gage, and standard time routing sheet for water outlet casting.

One each of the *A*, *B*, and *C* cores shown in Fig. 5–13 are made in the composite core box. The box is designed for a hand ram operation, which is the lowest-cost type of core box. No vents, screens, or blow plates are required. The

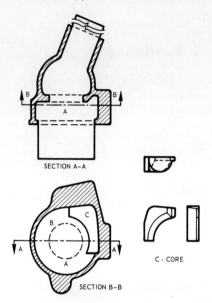

SECTION A–A

C - CORE

SECTION B–B

Fig. 5–13. Core assembly of water outlet casting of Fig. 5–11.

cores are carefully placed on a flat plate for drying, thereby eliminating the need for special driers. Also, it was estimated that the cost of making a few extra cores to replace those damaged in drying would be less than the cost of driers. The cost estimate for the core-making equipment is shown in Fig. 5–14; the total

CORE ESTIMATE SHEET

PART NAME: *Water Outlet*   DATE: *6/18/68*

PART NUMBER: *77230*   YEARLY VOLUME: *5,000*

TO CUSTOMERS DWG. DATED: *4/22/68*

TYPE OF CORE BOX: *Hand Ram - Composite* MATERIAL OF CORE BOX: *Alum.*

OPERATIONS	WOOD	CASTING	BENCH	MILL	LATHE	KELLAR	LAYOUT	TOTAL HOURS	MATERIAL	
CORE BOX **3** IN *Long*										
MASTER	30	12						42		
BOX			30	18			6	54	10.00	*Aluminum*
STEEL FACE										
VENT										
BLOW PLATE										
LAYOUT										
SETUP AND PROVING										
CORE ASSYS										
DRIERS	*None*									
PATTERNS	*n.a.*									
CAST	*n.a.*									
MACHINE										
SPOT FIT										
SPOTTING SLUGS										
MISCELLANEOUS										
DRILL FIXTURE										
							TOTALS	96	10.00	

LABOR ESTIMATES ARE IN HOURS *96 hrs.@ 2,62/hr. = $252.00*

MATERIAL IS IN DOLLARS

DETAILS: *One "A" Core; One "B" Core; One "C" Core*

Fig. 5–14. Cost estimate sheet for core making equipment.

cost of the equipment is $262. The cost of making 100 core assemblies is $1.5666.

Molding and Cleaning Costs. The estimated cost of molding and cleaning 100 water outlet castings is $.6898. The net hourly production is estimated as 87.2 molds. Fig. 5–15 is the routing sheet for the estimated cost of molding,

TEMPORARY OPERATION, EQUIPMENT, TOOL & GAUGE, & STANDARD TIME ROUTING

	ISSUE NO *1*	SHEET *1*	PART NAME *Water Outlet-Molding*		PART NO *77230*	
	FORGING NO	ROUGH WEIGHT *12.4/Mold*	MATERIAL *Gray Iron*		EFFECTIVE DATE *6/17/68*	
	CASTING NO	FINISH WEIGHT *1.686*	REFERENCE		GROUP NO	
BUDGET GROUP STD	CHANGE *Original*			CURRENT GROUP STD		
TOOL & GUAGES	BUDGET STD 1968	REMARKS	OPER	OPER. SEQUENCE & EQUIPMENT	STUDY	STD	NET HOURLY PROD	GROSS HOURLY PROD
Line No. - 4	*.428*	*221 PN*	5	*Mold-Cope & drag plates*		*$.428*	*Total 100 pcs.*	
Machine No.14 Pin Lift		*OR*		*Pos. per Mold -8 4 of 77230*	*100 Molds 1.71*		*87.2*	
Flask - 13x18				*and 4 of 27573*				
Drag-5in. Cope-6in.				*Operators -*				
Jolts -				*Chaplets*				
Oil -								
No. of job on Line	*$.285*		6	*Set cores 1 Man*	*Est $.214*		*Total / 100 pcs.*	
Line Speed -				*Cores- 4 assy. - See note*	*100 Molds 1.55*			
Note: 1-"A, 1-"B"				*Strainer -*				
and 1-"C" Core are	*.00397*	*221-32*		*all group operations*		*.00397*	*Total*	
pasted together to			10	*Pour molds, sand*	*22-1*	*.00368*		
make one assy.				*clamps, shakeout, pins*		*.000294*		
				and load baskets				
	.00131	*221-35*		*All Group Operations*		*.00131*	*Total*	
			15	*Vib. disc, gate port + chip*	*24-1*	*.00074*		
					8%	*.0000593*		
			16	*Barrel*	*24-1A*	*.00047*		
					8%	*.0000377*		
			17	*Load + unload 4 wheeler*	*24-3*			
	.00676	*221-36*		*all Group Operations*		*.00676*	*Total*	
			20	*anneal + Sandblast*				
				(2 men)	*5%*	*.000340*	*.2405*	
			30	*Grind*	*Est.*	*.000345*		
Pattern Layout		*29-45*	95	*Rectify (10% 120/Hr.)*		*.00396*	*.240*	
			100	*Inspect*	*D.W.*			
				Transfer to Shipping				
		894-17		*All Group Operations*	*D.W.*		*Total*	
			110	*Credit Group Operations*				
BOOK NOS *4 (6) (2)*				CHANGED STANDARD			TOOLS	

Fig. 5–15. Routing sheet for the water outlet casting cost estimate.

cleaning, and inspection of the water outlet casting. The cost of patterns for the cope and drag plates with four models of the casting is estimated at $464 for labor and $14 for material.

Manufacturing Cost. The manufacturing cost for the 5,000 castings is quoted as $1,687.82. Departmental overhead, general and administrative expense, and profit must be added to this cost to arrive at the price quoted to the customer.

ESTIMATING WELDING COSTS

The development of modern welding processes has provided a means of joining many different types of metal parts. Castings can be welded to castings, forgings to forgings, and forgings to castings, for example.

Effect of Welding Process Selection

All metals can be welded if the correct processes and equipment are used. Because the particular process chosen and means of applying it affect weld strength and costs, the welding estimator should acquaint himself with the various welding methods, especially those used in his plant.

Available processes range from the common gas or electric-arc fusion welding, to spot and seam resistance welding, to the relatively new electron beam, ultrasonic, and foil-seam butt welding. The methods of applying the welding process also vary, from hand-held torches, to welding machines, to semi-automatic and fully automated welding systems. Automatic welding should be considered only when either very high volume or rigid quality requirements exist. Although any welding process can be performed automatically, automatic welding requires more equipment, better fixturing, more setup time, and closer fitup. Study is necessary to determine whether better quality and overall cost savings will justify the higher costs of automatic welding.

Workpiece thickness and composition usually determine the type of welding process that can be used. For example, the inert gas-tungsten arc (TIG) process is economical for welding light-gage material, while the semiautomatic inert gas-metal arc (MIG) process is more economical for heavier gage materials and for nonferrous materials such as aluminum.

Process Plan and Tooling List

A process plan and tooling list for the process selected is needed before estimating can begin. The process plan, usually prepared by a process planner, lists each operation to be performed. In establishing operational sequence and tooling requirements for the welding operation, the process planner considers the following factors: (1) assembly configuration, (2) quantity, (3) delivery schedule, (4) quality requirements, (5) available equipment, and (6) personnel capabilities.

Cost Elements

The direct costs of manufacturing a welded part consist of material, labor, and tooling. Additional costs include quality control, packaging and shipping, and factory burden. Appropriate factors for overhead and profit are added to these costs to determine a final selling price.

Direct Material. Direct material includes all the material which becomes a part of the finished product; for example, the sheet stock, castings, stampings, forgings, or extrusions used in fabricating the product. The consumable electrode or weld wire used to provide additional metal in the weld groove is another major material item. Materials are also required for finishing treatments such as painting, porcelainizing, plastic coating, rubber bonding, metallizing, etc.

Scrap. The material costs of a welded product are affected by the scrap developed from the raw stock, weld rods or wires, standard component parts, and finishes. Any anticipated scrap should be considered in computing the basic material costs of each component or process.

Purchased Components. Purchased components such as rivets, bolts, nuts, latches, and hinges are often used on the welded products. The cost of such items (plus applicable transportation charges, purchasing burden, and handling costs) is added to the direct material portion of the estimate.

Electrodes. The consumable electrodes used to deposit filler metal in the weld groove represent a major cost item. Excessive consumption of electrodes (or welding rods) results from:

1) Failure to use optimum amount of the electrode. Discarding electrode stubs more than 2 in. in length is generally uneconomical.
2) Poor fitup. A poorly fitted joint requires more filler metal, and may increase electrode costs as much as 500 per cent.
3) Overwelding. Making more passes than necessary can quickly double the cost of electrodes.

For certain fusion welding processes, it is essential that an inert atmosphere surround the weld area. In some cases the welding rod has a coating which vaporizes during welding; in other cases a separate supply of inert gas must be used. Practices differ as to the method of estimating the cost of this gas; some companies include it as a part of material requirements, while others include it as overhead.

Direct Labor. Included in direct labor are the costs of all personnel working directly upon the fabricated part. The steps involved in producing a welded part include preparation, setup, welding, postwelding operations, postmechanical operations, and finishing.

Preparation. Raw stock must be prepared for welding. Preparatory operations include sizing, machining parts to print requirements, machining weld joints, and cleaning foreign material from the surfaces to be welded.

Setup. The setup for manual fusion welding, for example, includes assembling the pieces in the welding fixture, tack welding, operating the positioner, and/or preheating prior to welding. The additional step of adjusting and setting the time phase of the welding equipment is required for automatic fusion and resistance welding.

Welding. The labor used in actually making the weld is usually the largest labor cost component. However, *arc time*, or actual welding time, is a highly variable factor, ranging from a low of 10 per cent to a high of 75 per cent of the total time required to fabricate a welded product.

Factors influencing arc time include weld joint preparation, rate of weld deposit, type of welding process, and the number of passes required. The type of filler metal used also affects weld deposit time. For example, a mild steel electrode is usually deposited in two-thirds the time required for deposition of the same amount of stainless steel electrode, but it has a slower burn-off rate than iron powder electrodes.

Weld design, from the standpoint of accessibility, shop working and safety conditions, and the amount of equipment needed also influences time requirements. Fusion position, for example, is influential. Welding in the downhand position requires the least amount of time; overhead welding the most. Shop experience indicates a 10 per cent to 50 per cent longer weld time for horizontal welding over downhand welding while overhead welding may require 300 to 400 per cent more time than downhand welding.

Postwelding Operations. Manual or automatic fusion welds often require heat treating operations such as stress relieving, annealing, normalizing, hardening, and aging, and an appropriate amount must be added to the estimate for these operations. In the case of resistance welding, these operations are usually performed as part of the welding cycle, and are not estimated separately.

Postmechanical Operations. These include: (1) removal of excess weld metal, slag, and weld spatter, (2) rough or finish machining to dimensional requirements of the weldment, and (3) metal conditioning processes such as peening or roll planishing to reduce cracking tendencies physically and to increase weld bead strength.

Finishing. Brushing, burnishing, and other cleaning operations may be required, as may surface finishes such as painting, annodizing, or rubber coating. Each operation or process involves a labor cost.

Tooling. Special tools such as welding fixtures, machining fixtures, and machining templates are sometimes needed to complete the weld. Comparisons should be made between the cost of building tools in-plant and purchasing them from vendors.

Perishable tooling costs are generally charged directly to a specific part or product. However, durable tooling may instead be treated as a burden item. In cases of doubt, the estimator should seek a management decision.

Quality Control. Quality requirements for the welded part should be thoroughly investigated during the estimating process. Inspection and quality control are intended to assure quality at a predetermined level established by the customer and manufacturer. Meeting a high quality level may entail extensive inspection, welder testing, destructive and/or nondestructive part testing, and equipment certification. These items can add significantly to the cost of the job.

Inspection. Inspection affords a simple, economical, and accurate method of identifying trouble points and determining where to take corrective action to assure conformance to contractual requirements. Making an accurate estimate requires that the estimator be aware of the amount and level of inspection necessary.

One item that should always be inspected is fitup. Poor fitup can create gaps too wide to fill with the specified number of passes. Inspection prior to welding will prevent excessive corrective welding. Fig. 5–16 shows the effect of gap width on welding speed in relation to plate thickness.

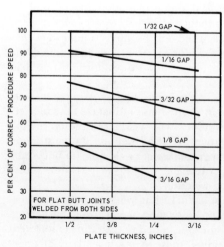

Fig. 5–16. Effect of gap width on welding speed of a downhand butt joint, square or grooved.

The degree and amount of inspection required depends upon the type of weld, i.e., commercial, low-stress, or aircraft and missile. The types of inspection can range from visual to detailed nondestructive and destructive testing programs.

Certification. Contracts for certain types of welding (such structural welds for buildings, bridges, ships, boilers, high-pressure lines, and military programs), often require that welders be tested or certified as to their ability to make satisfactory welds. Testing programs may range from an initial test to an extensive program involving continuous proficiency testing, and the company is often required to keep records indicating satisfactory completion of testing and certification requirements. The costs associated with such programs vary according to the degree of testing and the amount of required record-keeping.

Packaging and Shipping. The costs for packaging and shipping of the finished article depend on the type of packaging required, the distance the completed product will be shipped, and the method of transportation used.

Factory Burden. Burden includes plant space, lighting and heating and non-perishable tooling such as welding positioners and welding equipment. When special equipment must be purchased to fabricate a part, the estimator should seek a management decision regarding what charges to apply to the part and what portion to treat as capital investment.

Overhead and Profit

The completed cost estimate is generally forwarded to the accounting department for application of appropriate percentages for general and administrative costs (overhead) and profit.

Welding Estimate Example

To illustrate the foregoing costing procedure, the outlet cover shown in Fig. 5–17 will serve as an example. The outlet cover consists of 11 components

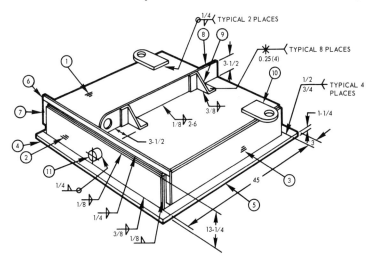

Fig. 5–17. Outlet cover fabricated as a weldment.

and is welded by two processes: manual arc welding and resistance spot welding.

Parameters. The parameters governing this example are: the order is for 100 units; an inspection level of 5 per cent is required; the units are to have a painted finish; machine or operator certification is required; factory personnel are well trained; and the necessary welding equipment is available. Because the customer has specified that he will pick up the finished covers, no special packaging will be needed.

Estimating Procedure. The steps involved in estimating the cost of fabricating the outlet cover are:

1) Calculate material requirements and develop total material costs ("Material Summary Sheet," Fig. 5-18)

2) Determine total engineering and tooling charges ("Engineering and Tooling Summary Sheet," Fig. 5-19)

3) List the welding processes necessary to assemble the outlet cover, and calculate deposited weld metal weight and welding time ("Welding Summary Sheet," Fig. 5-20)

4) List the labor operations (other than those required for welding and in-house tooling) and compute the time required for each ("Labor Summary Sheet," Fig. 5-21)

5) Summarize all costs and develop the total contract cost ("Cost Estimate Summary Sheet," Fig. 5-22).

Direct Material. The "Material Summary Sheet," shown in Fig. 5-18, reflects the cost of raw stock and other material used in the finished product.

Part Components. For components fabricated from raw stock, the estimator must analyze the part print to determine the dimensions of each component. From these dimensions he calculates the *volume* of each component, and multiplies the volume times the weight per pound of raw stock in order to determine the *weight* of each component. This weight is multiplied by the cost per pound of raw stock in order to obtain the *cost* of the component. The resulting cost figures for each component are shown in the right-hand column of the "Material Summary Sheet" (Fig. 5-18).

Such volume and weight calculations are not necessary for purchased parts. Note, for example, that the weight is omitted for Detail 11, a purchased pipe nipple, and that the cost of this item is simply listed under "Unit Cost," and the figure extended to the "Total Cost" column.

Finally, the total costs of the part components are added.

Scrap Factor. After determining the total cost of the part components, the estimator applies a scrap factor to allow for these losses. To keep scrap costs as low as possible, the estimator surveys raw stock vendors to determine whether material in the required sizes is commercially available. When required sizes are not available, raw stock must be purchased in larger units and cut to size, and a factor must be applied for raw stock cutoff losses. Allowance must also be made for material lossage due to machining and weld joint preparation.

The scrap factor of 1.5 used in this example was developed from previous shop experience and includes raw stock cutoff, as well as machining and weld joint preparation losses.

Weld Deposit. The amount of weld deposit (shown as "Total Electrode" on the Material Summary Sheet, Fig. 5-18) is determined from the "Welding Summary Sheet" (see Fig. 5-20).

Paint. The amount of paint needed is calculated on the basis of square feet of

COST ESTIMATE
MATERIAL SUMMARY SHEET

CUSTOMER _City Supply Company_   NO. _8302_
DATE _6/17/68_   ESTIMATOR _J.M.D._   SHEET _1_ OF _1_
PART NAME _Outlet Cover_   PART NO. _____

DET. NO.	DESCRIPTION	QUAN.	WEIGHT	UNIT COST	TOTAL COST
1	Cover Top	1	71 lbs	@ 11¢/#	$7.81
2	Cover End	2	76	@ 11¢/#	8.36
3	Cover Side	2	76	@ 11¢/#	8.36
4	End Flange	2	108	@ 11¢/#	11.88
5	Side Flange	2	96	@ 11¢/#	10.56
6	Top Bar	1	5	@ 11¢/#	.55
7	Side Bar	2	2	@ 11¢/#	.22
8	Rib	1	5	@ 11¢/#	.55
9	Gusset	4	4	@ 11¢/#	.44
10	Lift Lug	2	8	@ 11¢/#	.88
11	1½" Pipe Nipple	1		$1.25 ea.	1.25
	Total		451#		$50.86
	1.15 scrap factor		$50.86 × 1.15		58.49
	Total electrode		12.6#	@ 26¢/#	3.28
	Total paint		0.2 gal.	6.50/gal.	1.30
	Material Allowance				$63.07

Fig. 5–18. Material summary sheet.

COST ESTIMATE
ENGINEERING AND TOOLING

CUSTOMER _City Supply Company_   DATE _6/17/68_ NO. _8302_
PART NAME _Outlet Cover_ PART NO. _____   ESTIMATOR _JMD_ SHEET _1_ OF _1_

ITEM NO.	DESCRIPTION	BUY	MAKE	DELIVERY WEEKS	MAN HOURS	RATE	LABOR COST	MAT'L COST	AMORTIZED TOOL COST	NON AMORTIZED TOOL COST
1	Layout Template-8	✓		½						$68.00
2	Layout Template-9	✓		½						68.00
3	Layout Template-10	✓		½						68.00
4	Sub-Assy. Spotweld Fixture to locate 8 & 9	✓		1						140.00
5	Setup		✓		3.0	4.00	12.00		12.00	
6	Weldment Fixture-1,2, 3, 4 and 5	✓		1						288.00
7	Setup		✓		4.00	4.00	16.00		16.00	
8	Final Assy. Fixture + Sub Assy -6&7, -10 & 11	✓								211.00
9	Setup		✓		4.00	4.00	16.00		16.00	
									$44.00	$843.00

Fig. 5–19. Engineering and tooling summary sheet.

Fig. 5–20. Welding summary sheet.

surface area to be covered. For this example, it has been determined that one gallon of paint will cover 5 units.

Tooling. The "Engineering and Tooling Summary Sheet," shown in Fig. 5–19, covers the cost of the required fixtures, templates, and setups, as well as the labor costs for those items produced in-plant.

Fixtures and Templates. Fixtures and templates made in-plant require a separate estimate. (For the purpose of this example, it is assumed that any special tools will be purchased from an outside vendor. Bids are normally taken, and the best price received from a reliable vendor is applied to the estimate).

Welding fixtures should be designed to accommodate the welding methods that have already been chosen to be the most effective and economical for the particular job. The dimensional accuracy to which components have to be welded must also be taken into consideration. The fixtures used should be of no higher quality than necessary to obtain the required dimensional accuracy so that their cost can be held to a minimum.

Setups. Setup time includes the cost of labor required to assemble the fixture, check for the dimensional characteristics, and install the positioners.

Welding Operations. The "Welding Summary Sheet" shown in Fig. 5–20 is used to develop the costs directly related to the welding operations. To avoid overlooking any required welding, the estimator lists each part component on the summary sheet. Welding time and filler metal quantity is based on the size and length of the weld.

In calculating the weight of deposited metal, the first part component is

COST ESTIMATE
LABOR SUMMARY

CUSTOMER _City Supply Company_ NO. _8302_
DATE _6/17/68_ ESTIMATOR _J.M.D._ SHEET _1_ OF _1_
PART NAME_ Outlet Cover_ PART NO._____

DET. NO.	DESCRIPTION	QUAN.	SIZE	OPERATION	TIME—MIN
1	Cover Top	1	1/8 x 44 3/4 x 44 3/4	Shear	11.9
2	Cover End	2	1/4 x 12 x 44 1/2	Shear	18.4
3	Cover End	2	1/4 x 12 x 44 1/2	Shear	18.4
4	End Flange	2	1/4 x 3 x 51	Cut	5.0
5	Side Flange	2	1/4 x 3 x 45	Cut	5.0
				Machine	9.0
6	Top Bar	1	1/4 x 1 1/2 x 48	Shear	7.4
7	Side Bar	2	1/4 x 1 1/2 x 9	Shear	2.5
8	Rib	1	1/8 x 3 1/2 x 44	Shear	7.8
				Drill	.4
9	Gusset	4	1/8 x 7 1/2 x 8 (makes 2)	Shear	10.0
				Mark & Form	8.0
10	Lift Lug	2	1/4 x 6 x 9 1/2	Shear	4.2
				Drill	.8
				Total	108.8
					1.8 hr.
	Outlet Cover	1		Clean for paint	.4 hr.
				Paint	.3 hr.
				Total	2.5 hr.

Fig. 5–21. Labor summary sheet.

considered only as a starting piece because no joining weld is required. For the second piece, only the amount of welding needed to join it to the first piece is estimated and entered on the summary and, for each additional piece, only the amount of welding required to join it to the preceding parts is listed.

The weld lengths are entered in feet, and weld sizes are indicated by standard symbols with the weight per foot of each different weld size. Table V-18

Table V-18. Data for Fillet and 45-Deg. Bevel Welds.*

Weld Size (in.)	Weight (lb/ft)	Time (min/ft)
1/8	.027	1.5
1/4	.106	2.65
5/16	.166	4.15
3/8	.239	5.98
7/16	.325	7.40
1/2	.425	9.06
5/8	.664	13.10
3/4	.957	18.00
7/8	1.300	23.70
1	1.700	30.40
1 1/8	2.150	37.80
1 1/4	2.640	46.00

*Based on average 45 to 50 per cent arc time and welding deposition rates of 2.4 lb/hr for welds 3/8 in. and under; 3.6 lb/hr for 3/8 in. and over.

COST ESTIMATE
SUMMARY SHEET

CUSTOMER _City Supply Company_   NO. _8302_

DATE _6/17/68_   ESTIMATOR _JMA._ SHEET _1_ OF _1_

PART NAME _Outlet Cover_   PART NO. _____

MATERIAL ALLOWANCE _$63.07_   WASTE FACTOR _1.15_

LABOR ALLOWANCE _____   REJECTION FACTOR _1.03_

MACHINE & FINISH _.25 hr._   BURDEN FACTOR _2.15_

WELDING _.50 hr._   G & A FACTOR _1.07_

OFF STANDARD FACTOR _1.90_   PROFIT FACTOR _1.11_

TOOL & ENGINEERING ALLOWANCE ___ LABOR RATES ___

AMORTIZED _$44.00_   WELDER _$2.80/hr._

NON-AMORTIZED _$843.00_   MACHINE & FINISH _$2.75/hr._

TOOL DELIVERY _3½ weeks_

PART DELIVERY _7 per week beginning the fifth week._

QUANTITY	REJECTION ALLOWANCE	MATERIAL SELLING PRICE	LABOR SELLING PRICE	ENG'RG & TOOL SELLING PRICE	TOTAL SELLING PRICE
100	1.03	$77.16	$33.43 Mach.	$0.52	$184.12
			67.92 Weld		
			5.09 Insp.		
			$106.44 Total		

(1) 63.07 × 1.03 × 1.07 × 1.11 = 77.16

(2) .25 × 2.75 × 1.90 × 2.15 × 1.07 × 1.11 = 33.43
    .50 × 2.80 × 1.90 × 2.15 × 1.07 × 1.11 = 67.92
    7.5 × .05 × 2.80 × 1.90 × 2.15 × 1.07 × 1.11 = 5.09

(3) 44 ÷ 100 × 1.07 × 1.11 = $10.52

Cost 100 pieces $184.12 × 100   = $18,412.00

Non Amortized Tooling $843.00 × 1.07 × 1.11 = 1,001.23

Total Estimated Price $19,413.23

Fig. 5–22. Welding cost estimate summary sheet.

shows average arc time in minutes per foot and weight of deposited metal in pounds per foot for different sizes of fillet and 45-deg. bevel welds using mild steel electrodes. Deposited metal weight for different type welds may be calculated by multiplying the cross-sectional area of the weld by 12 in. The product (cu in/ft) is then multiplied by the weight of a cubic inch of weld metal, e.g., a deposited steel electrode weighs .238 lb/cu in. The result is weight per foot of deposited weld metal.

Manual Welding. For manual welding, the total weight of deposited weld metal is converted to an estimated weight of required electrodes by dividing by a factor of .55. This value is based on the assumption that the actual deposited weld metal is 55 per cent of the electrode's original weight.

Spot Welding. Estimates for spot welding are normally based on charts such as that shown in Table V-19. This shows typical point-closing to point-opening times for various thicknesses of material. These will vary slightly from one material to another and from one type of welder to another. The contact time for spot welding is taken as approximately 50 per cent of total weld time on thicker-gage materials. The remainder is required for fixture loading, etc. Contact time will be a slightly smaller percentage for thinner materials.

Table V-19. Data for Spot Welding.

Material Thickness (in.)	Time (sec.)
.025 to .025	1
.031 to .031	1
.042 to .042	1
.050 to .050	1
.062 to .062	1.5
.093 to .093	3
.125 to .125	4

Labor. The labor costs for operations other than in-plant tooling and welding are summarized on the "Labor Summary Sheet," shown in Fig. 5–21.

Machining and Cutting. To determine machining and cutting costs, the parts are itemized, and the time for each operation is calculated by using the company's standard accumulated data.

Painting and Plating. The cost of labor for painting and plating preparation is computed by use of standard time data.

Inspection. Inspection labor costs can be handled either as standard time data for each inspection operation, or as a percentage of the standard labor hours for manufacturing the part if a correlation can be developed. In this example a factor of 5 per cent of direct labor hours is used for inspection.

Total Contract Cost. The cost data and labor hours computed on the preceding summary sheets (Figs. 5–18, 5–19, 5–20, and 5–21) are transferred to the "Estimate Summary Sheet" (Fig. 5–22) preliminary to developing a total estimated selling price.

First, the estimator computes separate selling prices for: (1) material, (2) labor, and (3) amortized engineering and amortized engineering and tooling. This is done by applying the predetermined company rates and factors shown in the upper portion of Fig. 5–22 to the following formulas:

1) Material Selling Price = Material Allowance × Rejection Factor × G and A Factor × Profit Factor
2) Labor Selling Price = Labor Allowance × Labor Rate × Off-Standard Factor × Burden Rate × G and A Factor × Profit Factor
3) Amortized Engineering and Tooling Expense = Engineering and Tooling Allowance ÷ Quantity × G and A Factor × Profit Factor.

For the actual computations, see the lower portion of Fig. 5–22. The values determined by these formulas are added to arrive at the total selling price minus nonamortized engineering and tooling per unit. The unit price is multiplied by the desired quantity (100 in this case) to determine the total selling price (again excluding nonamortized engineering and tooling).

Next, nonamortized engineering and tooling must be determined. The formula used is:

Nonamortized Engineering and Tooling = Engineering and Tooling Allowance × G and A Factor × Profit Factor.

Total estimated contract price is then determined by adding the total selling price to the nonamortized engineering and tooling cost.

ESTIMATING FORGED PARTS

Forging, one of the oldest known methods of metalworking, is still widely used. Forged parts offer high strength and good fatigue-life characteristics.

Almost any forging can be made satisfactorily in one of several methods unless a definite restriction, such as grain flow position, eliminates one or more of the methods. The technique used depends largely upon individual shop practices and the type of equipment available.

The use of a drop hammer or of a forging press for a specific part, for example, often depends upon which machine is available because some types of forgings can be produced on either class of equipment. However, certain forgings are better suited to a drop hammer while others are better suited to the forging press. Generally, parts better suited to the forging machine cannot be made satisfactorily by either the drop hammer or the forging press.

Kinds of Forgings

Forgings are usually classified into four general groups according to their method of fabrication: (1) smith forgings, (2) drop forgings, (3) machine or upset forgings, and (4) press forgings. Smith forgings are also referred to as *open-die forgings*, while drop forgings, machine forgings, and press forgings are known as *closed-die forgings.*

Open-Die Forgings. Smith forgings[1] are classified as open-die forgings because flat dies are used to make them instead of impression dies. Open-die forgings are selected for large or irregularly-shaped parts that are impracticable to produce with impression dies or for short production runs where quantities are too small to justify the cost of impression dies.

Closed-Die Forgings. Drop forgings, machine forgings, and press forgings are classified as closed-die forgings because they are made by means of impression dies. The main advantages of closed-die forgings are: (1) lower operator skill requirements, (2) accelerated production rates, and (3) holding of closer tolerances. Because of the extra cost of the impression dies, closed-die forgings are more likely to be selected for long production runs.

Estimating Procedures

Until three or four decades ago, when the available forging materials and the shapes to be forged were considerably less complex than they are today, estimating procedures were likewise easier. Estimating departments were something new then and the work was usually done by the plant superintendent or by the sales department. One method of estimating forging costs was to use a chart showing pound prices for easy, medium, and difficult forgings, and for several grades of material, and multiply the net weight by a suitable pound price to obtain the estimated cost. Another method was to figure the cost of the forging stock needed and multiply its cost by a suitable factor. For example, if the forging stock cost $.50, it would be multiplied by a factor (perhaps $2\frac{1}{2}$) and the estimated cost would be $1.25. A third method was to use estimated man-hours multiplied by a material factor.

---

[1] Also known as "hand forgings," "hammered forgings," or "flat-die forgings."

Today, with many simple to highly complex materials, a greater variety in forging equipment, closer tolerances, metallurgical requirements and restrictions that were unknown several decades ago, thinner sections and smaller draft angles, and other new conditions, it has become necessary to develop detailed cost estimates in a consistent manner. The estimating procedures set forth in Chapter 4 are generally applicable to forgings; specialized procedures are outlined in the following sections.

Estimating Forms. Many standard shapes of forgings can be estimated from a price list after the weight has been calculated. If additional work is required, such as rough machining, testing, or special inspection, the cost is totaled on a simple estimate form.

Specially developed estimating forms are invaluable in detailing and presenting the various forging cost elements. Although commercial forging companies generally design their own forms to suit individual needs, most forms call for the same basic information (see Fig. 5-24).

Estimating Data. The availability of standard data from previous jobs greatly facilitates forging estimates. Helpful data include net and gross weights, production sequences, production rates, die life, scrappage rates, and tooling and die costs. Such data permit quick estimates on parts comparable to those produced previously.

In many cases, the new parts can be estimated by comparison with parts previously forged, making adjustments for slight variations in size, shape, material, quantity, etc.

When a forging is considerably different in shape or other specification from any of those previously produced, a basic study should be made which should include detailed sketches, or even a model, to work out all production details.

Cost Elements

Direct material, dies and special tooling, machining operations, scrappage, and labor are the most significant cost elements of the forging estimate. The cost of special items such as torch cutting, bending, and other auxiliary operations must also be computed. All of these costs must be analyzed carefully in preparing the detailed cost estimate.

Direct Material. Direct material is the largest single cost element for forged parts. In a typical cost estimate for a closed-die forging made of carbon or low-alloy steel in medium to high-production quantities, 40 to 60 per cent of the total factory cost is direct material when the forging is produced on a drop hammer or mechanical forging press. For parts made in a forging machine, the direct material percentage is sometimes higher.

Quantity. To determine direct material cost, the estimator first calculates the quantity of forging stock required for the desired production run, making appropriate allowances for material lossages.

Information from previous jobs permits quick computation of material allowances for comparable forgings. For example, filed data indicates that on a previous job a material allowance of 6.6 lbs. was required to fabricate a finished smith forging estimated to weigh 5.3 lbs. in its finished condition.

The estimator establishes a material allowance/finished weight ratio (6.6

lbs.:5.3 lbs. = 1.25:1) and applies this ratio to the finished weight of the casting being estimated. Thus, for a comparable smith forging with a finished weight of 7.2 lbs., the material allowance would be 9 lbs. (7.2 lbs. × 1.25).

When data from previous jobs are not available, the estimator must compute the material allowance. First, the *shape weight* (a theoretically correct weight as computed from engineering drawings or sketches) is calculated by dividing the part into suitable geometric sections and obtaining the volume of each section. The total shape weight is found by multiplying the total volume by the weight of the material per unit volume.

It is difficult to divide the part into suitable sections to obtain shape weights if the contours are comprised of irregular fillets, oddly shaped ellipsoidal sections, or other peculiar shapes. On such sections, the estimator should make an accurate layout of the section and use a planimeter to find the area of the irregular surface. This area is multiplied by the section thickness and material density to find the weight of the section. Table V-20 shows the density of metals commonly used for forgings.

Table V-20. Density of Metals (lb/cu in).

Metal	Density	Metal	Density
Aluminum	.098 to .101	Nickel alloys, A	.321
Beryllium	.067	Low carbon	.321
Beryllium copper	.298	Duranickel[2]	.298
Brass, forging	.305	Monel[3]	.319
Bronze, aluminum	.269 to .274	K Monel[2]	.306
Nickel aluminum	.273	Inconel[3]	.304
Tobin[1]	.304	Inconel X[3]	.298
Manganese	.282 to .302	Steel (cast)	.283
Bearing alloy	.292 to .299	Carbon	.284
High silicon	.308	Alloy	.284
Aluminum silicon	.278	18-8	.285
Phosphor	.311	13% Cr	.283
Copper	.325	14% W	.312
Duralumin	.102	22% W	.321
German silver	.306	Tantalum	.600
Lead	.409	Titanium	.160
Magnesium	.063	Tungsten	.700
Molybdenum	.368		

[1]TM Anaconda American Brass Company
[2]TM International Nickel Company, Inc.
[3]TM Huntington Alloy Products Division, The International Nickel Company, Inc.

Next, the *net weight* (the average actual weight of finished castings) is determined by multiplying the shape weight by a factor based upon data from previous shop experience. The net weight of most forgings exceeds the shape weight by 3 to 5 per cent.

Gross weight (the weight of the forging stock required to make a forging) is then determined. The simplest method is to multiply the net weight by a factor developed from previous shop experience. For closed-die forgings, this multiple ranges from 10 to 15 per cent. For open-die forgings, the multiple may be as high as 60 to 70 per cent.

When the appropriate multiple is not known, gross weight must be determined by adding to net weight the material loss due to factors such as: (1) flash, (2) scale, (3) tonghold, (4) sprue, and (5) cut waste.

First a *flash weight* loss is added to the net weight. (Flash is the excess metal extruded as a thin plate surrounding the forging at the die parting line to assure that all impressions are properly filled.) Flash is removed by shearing in a power press, either while the forging is still hot or after it has cooled. Average flash width and thickness varies with the weight of the forging, though hot-trimmed forgings have thicker flash sections than cold-trimmed forgings. Approximate flash dimensions for both types of forgings are shown in Table V-21.

Table V-21. Approximate Flash Thickness and Width on Forgings (1).

Net Weight (lbs.)	Thickness (in.)	Width (in.)	Weight of Flash (lb/in)
	Cold-Trimmed Forgings		
Up to 1	$^1/_{16}$	$^3/_4$	.0133
1 to 5	$^1/_{16}$	1	.0177
5 to 10	$^3/_{32}$	$1^1/_4$	.0333
10 to 15	$^1/_8$	$1^3/_8$	.0487
15 to 25	$^5/_{32}$	$1^1/_2$	.0668
25 to 50	$^3/_{16}$	$1^3/_4$	.0937
50 to 100	$^1/_4$	2	.1425
	Hot-Trimmed Forgings		
Up to 1	$^1/_8$	$^3/_4$	.0266
1 to 5	$^1/_8$	1	.0354
5 to 10	$^5/_{32}$	$1^1/_4$	.0553
10 to 15	$^3/_{16}$	$1^3/_8$	.0730
15 to 25	$^7/_{32}$	$1^1/_2$	.0941
25 to 50	$^1/_4$	$1^3/_4$	.1250
50 to 100	$^5/_{16}$	2	.1790
100 to 200	$^3/_8$	$2^1/_2$	.2670

Allowance must also be made for flash or punchout slugs from holes in the forging. Punchout slugs vary with the dimensions of the punched hole and the thickness of the section through which the hole is punched. In many cases, the punchout slug may be considered as having approximately the same thickness as the flash.

Scale is the material lost due to surface oxidation in heating and forging. The amount lost is a function of surface area, heating time, and type of material. Scale loss is generally computed as a percentage of forging net weight. For forgings under 10 lbs., 7.5 per cent of the net weight is added for scale loss; for forgings from 10 to 25 lbs., 6 per cent is added, and for forgings over 25 lbs., an addition of 5 per cent is a close approximation.

Scale loss is of greater importance as the surface area increases in relation to the shape weight, or where the sections are forged with very small flash allowance, as in press forging or upsetting work. Scale loss also increases with reheating.

The *tonghold* is a projection, $\frac{1}{2}$ in. to 1.0 in. long, at one end of the forging which is used to hold the forging. The *sprue* (the connection between the forging and the tonghold) must be strong enough to permit lifting the workpiece out of the impression without bending. Sprue loss, approximately 7.5 per cent of forging net weight, must be added to net weight. Some estimators provide for tonghold and sprue loss by use of a liberal flash allowance.

After gross weight is determined, an allowance must be made for *cut waste*. Cut waste consists of: (1) stock consumed as sawdust when bar stock is sized by saw cutting, and (2) bar end loss resulting from purchased bar length variations and short ends resulting from cutting the forging stock to an exact length. An allowance of 5 per cent of total gross weight is often used.

Cost. Direct material cost is calculated by multiplying total gross weight (plus cut waste allowance) by the cost per pound of forging stock. Direct material cost per forging is then obtained by dividing the resulting sum by the number of finished forged parts of acceptable quality.

Dies and Special Tooling. The cost of dies and tooling varies with the type of forging produced. These costs differ according to whether open-die or closed-die forgings are used.

Open-Die Forgings. As mentioned above, open dies are generally chosen for forgings made in short production runs or for large or irregularly-shaped parts.

Open-die forgings are produced without the use of impression dies, the part being forged between lower-cost flat dies and possibly being given some of its shape by the use of stock hand tools on the flat dies.

The cost and life of the dies and tooling are computed, with appropriate allowances for maintenance and repair, and the total cost is distributed evenly across the production run.

Closed-Die Forgings. Impression dies and special tooling are necessary for closed-die forgings. For small quantities, the cost of dies and other tooling can be a substantial factor.

The cost of impression dies depends upon the kind of forging to be made and upon the forging stock specified. A simple closed-die forging made from conventional material usually requires a set of dies and a set of trimming tools to shear away the flash.

As a general rule, as the cost of the forging stock increases because of higher alloy content, it forges with greater difficulty and the die impressions tend to wear faster, increasing the die factor cost increment. This rule applies to the copper-base alloys and to the aluminum-base alloys.

In some instances, it may be economical to use simple dies at a nominal cost and use a small amount of additional forging material because of a small number of forgings and low-cost material. With more expensive forging stock, it may be more economical to reduce the material used and add a forging step.

Once the general sequence has been established and the size of the forging unit determined, the size of the die blocks and the cost of the dies can be estimated. Small parts, where quantity permits, may be forged two, three, four, or six at a time.

The increment of die maintenance or die replacement, in closed-die forgings, is developed by estimating the total cost of a set of dies, including any resinkings, and dividing this total cost by the estimated number of pieces than can be produced from the dies. The cost and life of other tooling must be computed to develop a tooling replacement or repair factor. Data concerning die life and die and tooling costs from previous forging estimates assists in making these computations.

Machine Operations. Various types of forging equipment are available. The machine selected for a specific forging depends upon the size of the forged part, the availability of the machine, and economic factors related to machine operation.

Forging Equipment. Costs are developed for the various types of equipment by assigning an hourly rate to each of the production equipment items in the shop such as forging hammers, forging presses, forging machines, forge heating furnaces, trimming presses, and other equipment of a similar production nature. The hourly cost of operating a small drop hammer with its furnace and trimming press but without the production crew may be $25 an hour; one of the larger ones might cost $150 an hour. Much of the other supplementary equipment might be given an hourly rate. Usually, each forging plant develops its own machine-hour rates from its own experience through recognized cost accounting procedures. Table V-22 lists some representative machine-hour rates.

Table V-22. Typical Forge Shop Estimator's Chart.*

Machine	Rate per Hour					Setup Cost
	Machine	Furnace	Trim Press	Operators and Helpers	Total	
2,000-lb. board drop hammer	$ 15	$ 9	$ 3	$12	$ 39	$ 65
3,000-lb. board drop hammer	21	12	5	12	50	90
5,000-lb. board drop hammer	34	14	6	15	69	140
3,000-lb. steam drop hammer	36	14	5	16	71	140
6,000-lb. steam drop hammer	50	18	9	23	100	200
12,000-lb. steam drop hammer	75	23	12	35	145	400
20,000-lb. steam drop hammer	100	35	20	45	200	800
3-in. forging machine	20	10	....	15	45	150
6-in. forging machine	55	25	....	30	110	300
1,500-ton forging press	40	15	8	22	85	150
3,000-ton forging press	75	20	10	30	135	250
1,000-ton coining press	20	....	....	10	30	50
Snagging or grinding	6	....	....	4	10	....
Special inspection	4	....	....	5	9	....
Magnetic inspection	$ 6	$....	$....	$ 4	$ 10	$....

Shearing and steel handling	$2.50/gross cwt
Shipping and product handling	3.00/net cwt
Conditioning and inspection	1.00/net cwt
Normalize	1.00/net cwt
Anneal	2.00/net cwt
Heat treat (quench and temper)	$4.00/net cwt
General, administrative, and selling	9%

*The figures shown are fictitious and are given for purposes of illustration only. These figures are subject to occasional variation. An extra press or an extra man may be required for some types of forgings, and this would change the total.

Miscellaneous Operations. Not all the items of cost are on the basis of machine-hours. It has been found impractical to cover shearing, steel handling, cleaning, production material handling through the plant, shipping, heat treating, and certain other cost items on a machine-hour basis. These operations are generally estimated on a gross or net pound basis such as listed in Table V-22. These cost increments are usually smaller than the production increments but reflect a necessary part of the total cost.

Setups. Previously accumulated data are also used to establish setup costs for the various types of available forging equipment. See the applicable column in Table V-22.

Labor. Next the estimator adds labor costs to the estimate. Because the cost of labor varies with the type of machine used, it is convenient to summarize these costs as shown in Table V-22.

Scrappage Allowance. The expected allowance for spoilage or scrap is normally taken as a percentage of the factory cost. On small quantities, or on special compositions, the scrap percentage will be higher than for large production quantities.

Overhead and Profit

General and administrative costs (overhead) are usually calculated as a percentage of total manufacturing cost. If sales commissions or royalties are paid, they are added as separate items. The markup necessary to provide the desired profit is added to the total cost derived to arrive at the actual selling price.

Forging Estimate Example

A cost estimate for the gear blank shown in Fig. 5–23 has been requested in the production quantities of 1,000, 5,000, 10,000, and 20,000 forgings. The

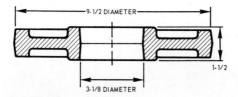

Fig. 5–23. Forged gear blank.

estimated life of each die sinking is 8,000 forgings, with a maximum die life of 40,000 forgings. The forging stock to be used is AISI C1045 steel. The forge shop operations will include pancaking (flattening), blocking, forging, trimming, and normalizing. A 3,000-ton forging press and 1,000-ton coining press are to be used to forge the blank. The numerous items comprising the estimate are recorded and summarized on the forging estimate form shown in Fig. 5–24.

Material Requirements. The gear blank can be conveniently divided into three sections for estimating the shape weight. Section 1 is the rim which is 9½ in. O.D. by 8¼ in. I.D. by 1¼ in. thick. Section 2 is a disk, 8¼ in. O.D. by 4½ in. I.D. by ½ in. thick. Section 3 is the hub, 4½ in. O.D. by 3⅛ in. I.D. by 1½ in. long. Table V-23 shows the calculations used to derive the shape weight,

FORGING ESTIMATE

CUSTOMER *S. R. V. Company*   DATE *6/17/68*

ADDRESS *Detroit, Mich.*

PART NAME *Gear Blank*   PART NO.

MATERIAL *C 1045*   SIZE *4" H.R.*   GR. WT. *21#*   NET WT. *16.0#*

MAT. COST	ITEM	WGT.	RATE	COST	ITEM	HRS.	RSK.	BLOCKS & MATERIAL	
BASE 7.00	C & H	21	2.50	.53	PLANE			2-11/4 φ ×4½ } 900#	
GRADE .50	SHIP.	16	3.00	.48	TURN	80	30	2-16 φ ×5 } @ .70	.030 —
SIZE .50	CLEAN	"	1.00	.16	TEMP.	12		Trim Mat. 400# @ .30	120.—
FGHT. .15	COND.	"	1.00	.16	SINK			192 hrs @ 10.00	1920.—
	NORM.	"	1.00	.16	DOWEL			Total dies	$2670.—
	ANN.				TRIMS.	90	30	RSKS	
	H.T.				COMB.			68 hrs. @ 10.00	680.—
	BRIN.							Die life 8000 pcs	
					MISC.	10	8		
	D.M.			.38					
TOTAL 8.15	TOT. FIXED COST			1.87	TOTAL	192	68	DIE COST $2670	

QUANTITY					1000	5000	10000	20000
PRODUCTION	*pcs./hr.*				100	150	170	190
UNITS			OP.	HLP.	RATE			
SETUP	3000 J Press				250.00	.250	.050	.025	.013
"	Coin Press				50.00	.050	.010	.005	.003
FORGE	" "		1	4	135.00	1.350	.900	.795	.710
"	✓								
TRIMS (H) (C)									
	Coin - 250/hr.		1	1	30.00	.120	.120	.120	.120

TOTAL FIXED COST		1.870	1.870	1.870	1.870
MATERIAL	21 × 1.05 × 0.0815	1.800	1.800	1.800	1.800
SUB-TOTAL		5.440	4.750	4.615	4.516
SPOILAGE	3 %	.16	.14	.14	.134
FACTORY COST		5.60	4.89	4.75	4.65
SALES & ADMINISTRATION	9%	.50	.44	.43	.42
COMM. OR ROYALTY					
TOTAL COST		6.10	5.33	5.18	5.07
MARKUP OR PROFIT	10%	.61	.53	.52	.51
QUOTING PRICE		6.76	5.86	5.70	5.58
LB. PRICE		.42	.37	.36	.35

SEQUENCE	REMARKS
Pancake	
Block	
Forge	
Comb. Trim	
Normalize	

Fig. 5–24. Forging estimate for gear blank of Fig. 5–23.

net weight, and gross weight of the forging. As shown, net weight is usually 3 to 5 per cent greater than shape weight.

To the net weight is added an allowance for flash around the outside of the blank and within the hub, scale loss and, for most forgings, a tonghold and a sprue between the forging and the tonghold. However, for this gear blank, tonghold and sprue are not required. Table V-21 lists typical flash allowances for hot- and cold-trimmed forgings.

Material Costs. The material costs shown on the cost estimate in Fig. 5-24 include the mill price plus extra charges for grade, size, and freight. Not shown on the estimate, but often applicable, is a "quantity extra." For small quantities not available from the mill, the forger may often be required to purchase the metal from a warehouse at a higher price which is called a "quantity extra."

Table V-23. Calculation for Weight of Gear Blank in Fig. 5–23.

Section 1:
  $9\frac{1}{2}$ in. diameter = 20.08 lb/in

  $8\frac{1}{4}$ in. diameter = $\underline{15.15}$
               4.93 lb/in × $1\frac{1}{4}$ in. = 6.17 lbs.

Section 2:
  $8\frac{1}{4}$ in. diameter = 15.15 lb/in

  $4\frac{1}{2}$ in. diameter = $\underline{4.51}$
              10.64 lb/in × $\frac{1}{2}$ in. = 5.32 lbs.

Section 3:
  $4\frac{1}{2}$ in. diameter = 4.51 lb/in

  $3\frac{1}{8}$ in. diameter = $\underline{2.17}$
               2.34 lb/in × $1\frac{1}{2}$ in. = $\underline{3.51}$ lbs.
Shape weight   15.00 lbs.
Plus 5% shape allowance   .75
    Net weight   15.75 lbs.
Flash allowance:
  From Table V-22, for a 16-lb. hot-trimmed forging, flash width = $1\frac{1}{2}$ in.; $\frac{7}{32}$ in. thickness; weight per inch = .0941 lb.
  For a ring, the center of gravity is .707 times the width from the I.D. The weight of the outer ring of flash is:
  2(.707 × 1.5) + 7.5 = 11.6 in. diameter
  11.6π × .0941 = 3.42 lbs.
  The slug in the hub is 2.173 lb/in for $3\frac{1}{8}$ in. diameter × $\frac{7}{32}$ in. = .48 lb.
  Scale loss: .075 × 15.75 = 1.18 lbs.
  Gross weight: 15.75 + 3.42 + .48 + 1.18 = 20.83 lbs. or 21 lbs.

Fixed Costs. These estimated costs are based on the forging weight. The charges for each operation per hundredweight are shown in Table V-22.

Die Costs. Tooling costs are shown in the "Blocks & Material" column at the top of the estimating form (Fig. 5–24). The basic die and tooling cost, as shown, is $2,670. Because die life is estimated at 8,000 forgings, $2,670 is the total die cost for the 1,000 and 5,000 production quantities.

After 8,000 forgings, additional labor is required to machine the forging and trimming dies, resink the forging die, and recondition the trimming die so that they will produce forgings within the required tolerances. The time required for this series of operations (68 hrs.) is multiplied by the hourly rate ($10) to arrive at a cost of $680. For 10,000 forgings, the cost of one machining, resinking, and reconditioning operation (or $680) is added, making a total of $3,350. For 20,000 forgings, another series of operations is required, making a total die cost of $4,030. The tooling cost is a separate item on the quotation and is not included in the quoted forging price.

Machine Costs. Examples of these costs are given in Table V-22. On the estimate form (Fig. 5–24) the setup costs and production costs are prorated per piece.

Quoting Price. The costs recorded on the upper portion of the form are summarized on the lower portion to determine the quoting price. When calculating the material cost, a 5 per cent scrap allowance is made. Three per cent is added for spoilage, 9 per cent for sales and administration costs, and 10 per cent for profit. As mentioned above, the quoted price does not include the tooling cost.

ESTIMATING METAL STAMPING COSTS

Assume that the sales department has requested a cost estimate for the stamped lever shown in Fig. 5–25. The customer's initial order is for 500 parts, to be followed by an order for 5,000 parts per month for one year, for a total of 60,500 parts.

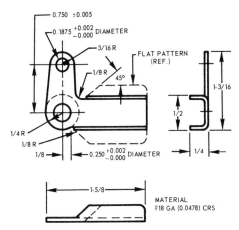

Fig. 5–25.   Lever made by pressworking operations.

Process Plan

The cost estimate request should be accompanied by a process plan specifying the stamping process to be used. When the process plan is not included, the estimator either requests this item from the process planning department or develops the plan himself. In either case, the person preparing the process plan must analyze the part design and be aware of the desired production quantity. Data available from previous stamping jobs greatly facilitates selecting the most efficient and economical processes and methods available in the plant.

Because no process plan was provided in this case, the estimator must determine the most economical method of manufacturing the lever to specifications. Three stamping processes commonly used in the plant are considered:
1) Pierce and blank in a compound die from hand-fed stock and form in a second die.
2) Blank, pierce, and form with three separate dies from hand-fed strip stock.
3) Pierce and blank in a compound die from coil stock and form in automatically fed forming die.

After evaluating each of these processes, the estimator selects the first as the most efficient and economical.

The second process was examined in detail but rejected because its slight reduction in material cost does not offset the increased cost of using three separate dies. Although the company owns automatic feeding equipment, the estimator determines that the third process would not be economical for production quantities less than 100,000 parts.

Estimating Procedure

After carefully analyzing the selected process, the estimator calculates the manufacturing cost for the stamped part. First, the cost of material is computed in the top portion of the estimating form (Fig. 5–26). Next, die cost, die repair, die set, labor, and burden costs are calculated in the "Operations" portion of

MATERIAL: .0478 ± .004 C.R. Steel	WT. EACH	PRICE CWT.	COST PER 'C'	SCRAP CREDIT	WT EACH	PRICE CWT.	VALUE PER 'C'	NET MAT COST
.0478 x 1⅞ x 41½ x .283	.0317	6.45	.205	15%	.005	120	.005	.20
32 Reverse Stock								

CUSTOMER J.R.U.Co.

OPERATIONS	DIE COST	MACH	HELP	LABOR	BUR DEN	TOTAL L & B	PCS PER HR	COST PER'C'	DIE REPAIR	DIE SET	PARTIAL MINOR COST	
Compound	600	5A	1F	2.05	2.70	4.75	2400	.20	.09	.07	SHEARING 104 cwt	.015
Form	300	5A	1F	2.05	2.70	4.75	1600	.30	.04	.07	STEEL HANDLING	.005
											INSPECTION	.10
											SALVAGE	.005
											TUMBLE	—
											WASH	—
											FLOOR HELP	—
											SUPERVISION	—
											SHIPPING	.005
	50										PARTIAL MIN. COSTS	.13
	50										MAJOR COST	.50
	1000						.50	.13	.14		CONVERSION L & B	.63

	QUAN.	500	5M	Total	60M /YR	MATERIAL COST PER CWT.				
	DIE REPAIR	200	.13			BASE				
500 Initial Order	DIE SET	280	.28			FRT.				
	CONV L & B	.43	.63			GAGE WIDTH				
	TOTAL L & B	5.90	1.04			LENGTH				
5000/ month	METAL	.20	.20			P & O				
60M/ year	% SPOILAGE	.12	.02			QUAL.				
						PKG.				
							6.45			
	CONTAINERS									
	FREIGHT									
	TOTAL					TOTAL	6.45			
	% ADM					DATE				
	COST PER "C"	6.20	1.26	% Adm.						
	QUOTE									
	PRELIM CHARGE									

REMARKS

Fig. 5–26. Cost estimate sheet for the stamping shown in Fig. 5–25 using two dies and strip layout (Fig. 5–27a).

the form, and appropriate amounts are entered in the "Partial Minor Costs" section of the form for miscellaneous operations such as shearing, handling, inspection, salvage, and shipping. The labor and burden costs are then added to "Partial Minor Costs" to arrive at "Conversion L & B" (labor and burden).

All costs are summarized in the bottom left portion of the form to develop a cost per hundred pieces. First, die repair, die set, and "Conversion L & B" are added to arrive at "Total L & B." To this cost, the estimator adds the material cost and an allowance for spoilage to obtain a manufacturing cost per hundred pieces.

Finally, the estimator sends the completed estimate to the accounting or sales department to calculate selling price.

Material

Material is one of the most important considerations in producing stamped parts. In checking engineering specifications and the process plan, the estimator pays close attention to material requirements, suggesting alternative materials to effect cost reductions when possible without loss in product quality or functionality. The estimator also considers the form in which the material is available, e.g., sheets, strips, etc., because the stamping process selected determines the form that must be used. For example, strip stock must be used for automatic equipment.

Stock Utilization. When the material has been determined, the estimator considers stock utilization. He studies the positioning of the blanks on the coil or strip and selects the layout that produces maximum stock usage and minimum scrap. This procedure, sometimes called "blank nesting," requires careful and intricate planning. Data from past estimates are seldom of any assistance because of the variations in part shape and stock size.

If bending or forming operations are required, the grain direction of the stock may be of importance. To aid in positioning the part on the strip, the estimator makes a template of the blank pattern of the part. Since there are flanges on the part, the bend lines cannot be more than 45 deg. to the grain direction of the metal. In this case, the grain direction is parallel to the edge of the strip or coil. Fig. 5–27 shows three possible strip layouts. Views *a* and *b* represent layouts for strips 41.5 in. long which are fed through the die twice. View *c* represents a layout for a strip 41.5 in. long which is fed through the die once, or for coil stock fed by an automatic stock feeder.

The strip layout shown in Fig. 5–27*a* allows 32 blanks to be cut from each strip of stock. The strip layout shown in Fig. 5–27*b* provides 40 blanks per strip. Each strip is fed through the die at the feeding increment indicated, then reversed and fed through the second time. The weight of stock per blank for the layout of View *a* is .0317 lb.; for View *b*, .033 lb.; for View *c*, .0304 lb. When using coil stock, the weight per piece for the strip layout of View *c* is .0296 lb. since there is less waste stock at the end of each coil than for strip stock. Computations for determining material weight are made in the "Material" section of the cost estimate sheet (see Fig. 5–26 for computations for View *a*).

The use of the layout shown in View *c* would result in a lower material cost, but this layout is rejected because of the extra cost of automatically-fed forming dies. The layout shown in View *a* is selected as being best for the hand-fed process selected.

Material Cost. Material cost is computed in the "Material" section of the estimate form (Fig. 5–26). To determine material cost per hundred parts, the weight per hundred pieces (3.17 lbs.) is multiplied by the material cost per pound ($6.45/cwt or $.0645/lb), and the resulting cost ($.205) is modified by the subtraction of an allowance for scrap.

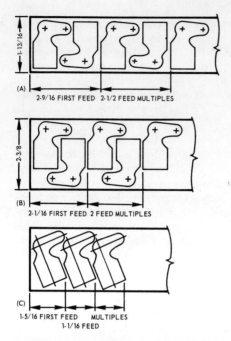

(A)

2-9/16 FIRST FEED   2-1/2 FEED MULTIPLES

(B)

2-1/16 FIRST FEED   2 FEED MULTIPLES

(C)

1-5/16 FIRST FEED   MULTIPLES
1-1/16 FEED

Fig. 5–27. Three strip layouts for stamping of Fig. 5–25.

To determine the scrap allowance, weight per piece (.0317 lb.) is multiplied by the scrap credit factor of 15 per cent. The scrap weight per 100 pieces (.50 lb.) is multiplied by scrap value per pound ($.012/lb) to obtain the scrap allowance per hundred pieces which is $.006 (rounded off in the "Value per 'C'" column to $.005).

"Net Material Cost" is determined next by subtracting scrap credit per hundred ($.005) from material cost per hundred ($.205), and the resulting cost of $.20 is entered in the appropriate column.

Machine Operations

In the "Operations" section of the estimate form (Fig. 5–26), the estimator enters the cost of the dies including tryout and engineering costs, computes the labor and burden costs associated with the stamping machine operations, and enters amounts for die repair and die sets.

Die Cost. The cost of the compound and forming dies, and the tryout and engineering costs associated with them are entered and totaled as $1,000. These are permanent dies, and because their cost is nonrecurring, it remains the same regardless of the production quantity.

Labor and Burden. Under the appropriate columns, the estimator enters hourly costs for labor and burden and adds these under the "Total L & B" column. These costs are then divided by the production rate per hour and the resulting figure multiplied by 100 to arrive at a labor and burden "Cost per 'C'."

Die Repair and Die Set. This is the cost of repairing and resetting the compound and forming dies. These costs are entered for both operations and totaled.

Minor Costs

Next the estimator computes the costs of the labor and burden involved in miscellaneous operations such as shearing, steel handling, inspection, salvage, shipping, etc. The cost of each of these items is determined from a study of data developed from previous stamping operations and entered in the appropriate place under the section entitled "Partial Minor Costs."

Conversion Labor and Burden

This is determined by adding the labor and burden cost from the "Operations" section and the "Partial Minor Costs" section. Conversion labor and burden in this case is $.63.

Total Manufacturing Cost

The total manufacturing cost for the stamped part is determined by adding all the applicable costs in the lower left section of the estimate form. An additional amount is included for spoilage. The total cost for a production lot of 500 pieces is $6.20 per hundred; for production lots of 5,000 pieces the cost per hundred decreases to $1.26.

Quotation

The price quoted to the customer is determined by adding percentages for general and administrative costs (overhead) and for profit. Assuming an overhead factor of 10 per cent of total manufacturing cost and a profit factor of 15 per cent, the computations necessary to develop a quotation may be summarized as follows:

Total Manufacturing Cost per Hundred	$1.26
Overhead (10% of $1.26)	.13
Profit (15% of $1.26)	.19
Total Cost per Hundred	$1.58

The quotation for 65,000 pieces (initial order of 500 plus 5,000 pieces per month for one year) would be $1,027 plus $1,000 for dies.

ESTIMATING PLASTIC PARTS

To estimate the cost of manufacturing plastic parts, the estimator follows the basic steps enumerated in Chapter 4. This example furnishes a general orientation to plastic parts estimating, and a specific application is included for a molded plastic tank (Fig. 5–29).

Fabricating Processes

The processes available for fabricating thermoplastic products include: injection molding, blow molding, thermoforming, slush molding, extrusion, and casting. For thermosetting materials, the available processes include: compression molding, transfer molding, matched metal molding, premix molding, pressure laminating, filament molding, spray molding, and casting. These are commonly used techniques; special shapes and problems often necessitate variations.

Process Planning

Ordinarily, the cost estimate request is accompanied by a process plan. When the process plan is not provided, the estimator requests this item from the process planning department or prepares it himself.

Fabricating Methods. The person preparing the process plan must be aware of plant capabilities and equipment. In any given plant, the choice of molding methods may be limited by the available equipment. Where a variety of equipment is available, the estimator may be responsible for selecting the most suitable and economical process. Each fabricating method has advantages that may make it the proper choice for a particular product. Tooling costs, cycle time, labor, and overhead vary according to the method selected.

Part Analysis. The person preparing the process plan works from a drawing of the part or other specifications received from the customer. The drawing or specifications should establish dimensional tolerances; surface condition; color; and the mechanical, chemical, electrical, or thermal properties of the material or product.

The process planner reviews only the general design characteristics of the part, not the physical and chemical stresses to which the part will be subjected once placed in use by the customer. The estimator, whether or not he prepares the process plan, should note dimensional tolerances because these directly affect labor and tooling costs.

Plastic Fabrication Practices. The part design should be reviewed for good plastics fabrication practices such as:

1) Generous use of fillets because square corners concentrate stresses and increase mold costs
2) Use of thin cross sections to speed up cycle time
3) Use of stiffening ribs on large flat surfaces
4) Holes through the section wherever possible which permit support of the core pin in both halves of the mold (blind holes should not be more than 2D deep)
5) Hole spacing sufficient to permit the material to flow smoothly
6) Draft allowances where necessary to facilitate removal of the part from the mold.

Labor and Burden Costs

Labor and burden costs for both major and secondary operations are assigned on the basis of machine operation time. The costs per piece may be reduced by having the operator tend more than one machine and/or by using multiple cavity dies.

Cycle Times. The molding cycle times for thermosetting plastics are determined by curing rate and section thickness. Preheating of the thermosetting plastics reduces the cycle time. Unless kept within prescribed tolerances, a change in moisture content of the material will result in over- or undercuring.

Secondary Operations. After fabrication, it may be necessary to trim, drill, tap, or otherwise finish a plastic part.

Filing. Flash, sprue, and gate marks are usually removed by filing or by abrasive methods.

Drilling. Plastics should be drilled with a 60- to 90-deg. included angle point drill with highly polished flutes. Speeds of 100 to 300 sfpm. are used for most plastics. The speed may be reduced to about 75 sfpm. for plastics with abrasive fillers. Holes may be tapped at 40 to 50 sfpm. Water is usually satisfactory as a coolant. Holes with threads larger than $\frac{1}{4}$ in. diameter should be molded. Threaded inserts should be considered for additional strength.

Turning. Feed rate and depth of cut for turning operations are a function of the material's physical properties. Turning speeds using high-speed steel tools range from 200 to 600 sfpm.; for carbide tools, the speed ranges from 500 to 1,500 sfpm.

Milling. The speed for milling with high-speed steel cutters is approximately 400 sfpm.; for carbide cutters, the range is 1,200 to 1,600 sfpm.

Material

The estimator is concerned with the moldability and fabrication characteristics of the plastic specified. He must know if the plastic is thermoplastic or thermosetting, for example, because fabricating procedures and tool costs are affected.

Part Weight. The volume or weight of the finished part must be known to estimate material requirements. This may be determined by one of three ways:

1) Weighing a sample of the part being estimated
2) Immersing a model of the part in water and using the weight of the displaced water to calculate the weight or volume of the finished part
3) Dividing the part into basic geometric forms and calculating the volume.

Scrap Allowance. Depending upon the fabricating process used, an allowance for shrinkage, flash, trim stock, process loss, sprues, and runners is added to the weight of the finished part. This allowance is usually expressed as a percentage of the finished weight, and determined from past performance.

Direct Material Cost. Material cost is determined by multiplying the cost per pound of raw material by the finished part weight after an allowance has been made for scrap.

Indirect Material

Indirect materials such as release agents, mold cleaners, vacuum bleeder liquids, and packaging supplies are usually considered as overhead. This simplifies the estimating procedure, and is generally sufficiently accurate to distribute these costs.

Tooling Costs

Tooling costs vary with the process to be used. In the manual layup of laminated plastics, a form block must be made. The cost of these blocks vary with their size and complexity. Pressure laminating may require the construction of a die similar to that for metal forming.

The cost of injection, compression, and transfer molds varies with the complexity of the part, the number of cavities in the mold, and whether the cavities are in inserts or machined directly into the die block. The transfer molding of thermosetting materials permits the use of thinner die sections and increased die life over compression molding.

Cost Estimate Example

The estimate request form shown in Fig. 5–28 was filled in by the sales department and sent to the estimating department. Because part drawings were not supplied by the customer, the estimator drew the sketch shown in Fig. 5–29

ESTIMATE REQUEST FORM

CUSTOMER *X Y Z Company*   NO. *6199*
SALESMAN *W. J. Able*   DATE *6/17/68*
DATE RECEIVED *6/18/68* DATE DUE *7/3/68* DATE TO ESTIMATING *6/19/68*
DESCRIPTION OF PRODUCT *Tank - approximately 36"x15"x25" with 1/2"*
wide exterior flange on open end. Thickness -
0.100 ± 0.015, 2° draft OK on all sides.

QUANTITY *100*
DELIVERY REQUIREMENTS *Must have all parts by end of September,*
1968. Customer will pick up, unboxed in lots of 25.

PRODUCT REQUIREMENTS	YES	NO	DESCRIBE
SPECIAL CHEMICAL RESISTANCE PROPERTIES		✓	
THERMAL RESISTANCE PROPERTIES		✓	
ELECTRICAL PROPERTIES		✓	
FIRE RESISTANCE PROPERTIES		✓	
OTHER	✓		*Exterior surface ready for prime and paint.*

DIMENSIONAL REQUIREMENTS (THICKNESS TOLERANCES, CRITICAL DIMENSIONS, ETC.)
Commercial tolerances - customer will furnish mfg.
drawings for requote prior to purchase.
LIST OF SPECS. & DRGS. CALLED OUT BY CUSTOMER *None*

FINISH REQUIREMENTS *Customer expects some minor pinholing in*
exterior surfaces. Sandblasted surfaces OK., interior surfaces
not critical regarding surface condition. Pigmented gelcoat
on exterior surfaces (light grey). (Customer will finish with
light grey primer and enamel).

Fig. 5–28.   Cost estimate request form (2).

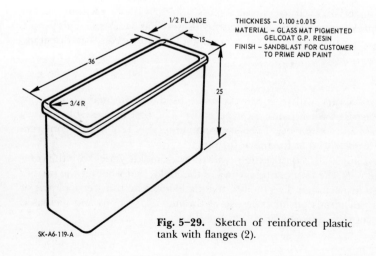

1/2 FLANGE

15

36

3/4 R

25

THICKNESS – 0.100 ±0.015
MATERIAL – GLASS MAT PIGMENTED
          GELCOAT G.P. RESIN
FINISH – SANDBLAST FOR CUSTOMER
         TO PRIME AND PAINT

SK-A6-119-A

Fig. 5–29.   Sketch of reinforced plastic tank with flanges (2).

for guidance in his estimate. The drawing assists in visualizing possible manufacturing difficulties such as sharp radii, reverse flanges, negative draft, and inaccessibility of some areas of the part to the workman. These difficulties affect labor costs, and should be called to the customer's attention for appropriate design modifications.

Material Requirements

The calculations for determining material requirements are shown in Fig. 5–30. The form shown in Fig. 5–31 includes a checklist of the materials the estimator must consider when making the estimate. The volume of the laminate and gelcoat were computed. Using the density of each, the total weight of the tank was calculated. The estimator also calculated the waste factors for the glass

COST ESTIMATING CALCULATION FORM

CUSTOMER XYZ Company NO. 6199
DATE 6/17/68 ESTIMATOR W.A.K. SHEET 1 OF 1

SHOW ALL CALCULATIONS

Ref. dwg A6-199-A

AREAS Flange 2(15+36)½ = 51 sq. in.
sides 2(15 x 25) + 2(36 x 25) = 2550
bottom 15 x 36 = 540
3141 sq. in.
21.8 sq. ft.

TRIM PERIMETER (37 + 16)2 = 106 in.

THICKNESS 0.100 in. = 0.090 laminate + 0.010 gelcoat

VOLUME laminate 3141 x 0.090 = 282.7 cu. in.
gelcoat 3141 x 0.010 = 31.4
314.1 cu. in.

DENSITY laminate 0.05563 lb/cu. in. (51% glass, 49% resin
gelcoat 0.05054 lb./cu. in.
Total 0.05563 x 0.9 + 0.05054 x 0.1 = 0.0551

WEIGHT (282.7 x 0.05563) + (31.4 x 0.05054) = 15.7 + 1.6 = 17.3 lb.

WASTE FACTORS

GLASS (4 plies of 1.5 oz. mat 38" wide — resin & gelcoat

2 pcs. 90 x 19 = 3420 sq. in. — $W_{LAM} = 1.11 \times \frac{\Delta + 1}{\Delta}$
2 pcs. 70 x 38 = 5320
4 pcs. 28 x 19 = 2128 — $\Delta = \frac{3141}{106} = 29.6$ sq. in./in.
4 pcs. 28 x 38 = 4256
total 15124 sq. in. — $\frac{\Delta + 1}{\Delta} = 1.034$

$W_{LAM} = 1.034 \times 1.11 = 1.148$ lb.

gross glass weight: $\frac{15124}{144} \times \frac{1.5}{16} = 9.85$ lb.

net glass weight 15.7 x 0.51 = 8.01 lb.

$W_E = \frac{9.85}{8.01} = 1.23$

PVA bag: 14 ft. of 3 mil by 52 in. sheet x ½ lb./ft. = 1.17 lb.

Fig. 5–30. Calculation sheet for plastic tank of Fig. 5–29 (2).

COST ESTIMATING
BILL OF MATERIAL

CUSTOMER _XYZ Company_   NO. _6199_

DATE _6/17/68_   ESTIMATOR _W.Q.K._   SHEET _1_ OF _1_

1	CLOTH		8	FINISH		15	CATALYST		✓
2	MAT & ROVING	✓	9	HARDWARE		16	PROMOTED		✓
3	GELCOAT	✓	10	PVA		17	PIGMENTED	✓	
4	RESIN	✓	11	VENDOR		18			
5	CORE		12	PACKAGING					
6	ADHESIVE		13	BLEEDER					
7	FILLER		14	BINDER					

CODE	MATERIAL DESCRIPTION	NET AMOUNT	WASTE FACTOR	GROSS AMOUNT	MAT'L COST/LB.	COST
2	1.5 oz. mat. (38 in. wide)	8.01#	1.23	9.85#	$0.60	5.91
3	light grey gelcoat	1.60#	1.148	1.84#	0.49	0.90
4	general-purpose resin	7.69#	1.148	8.83	0.345	3.05
10	14 ft. of 3 mil x 52 in. PVA	-	-	1.17#	1.70	1.99
15	MEK peroxide (2%)	0.15#	1.148	0.17#	1.80	0.31
16	cobalt naphthenate (1%)	0.08#	1.148	0.09	0.53	0.05
		18.17#		21.95#		12.21

Fig. 5–31.   Material estimate sheet (2).

cloth, resin, and gelcoat. These factors were used in estimating the gross material requirements listed in Fig. 5–31.

Labor

The time required to fabricate the laminated plastic tank is shown in Fig. 5–32. A labor operations checklist is provided to remind the estimator to include all the necessary items. The labor operations are described in sequence, and the number of men, time in minutes, and man-minutes for each operation are listed.

Tooling

The tools required to make the part and the data pertaining to their design and manufacture are shown in Fig. 5–33. This form does not provide spaces for itemizing the labor and materials required for each tool. If this cost data is required, the forms shown in Figs. 5–31 and 5–32 may be used.

Cost Estimate Summary

A summary of the cost estimate is shown in Fig. 5–34. At the top of the form are spaces for summarizing the dimensional data compiled during the initial calculations and other important processing information. Spaces are also provided for listing the prices for different quantities should the customer request this type of estimate.

ESTIMATING TUMBLING AND VIBRATORY FINISHING COSTS

A wide range of surface improvement operations are economically possible through the use of tumbling and vibratory machines. Deburring, cleaning, polishing, mirror finishing, and honing processes are accomplished by proper

COST ESTIMATING
LABOR REQUIREMENTS

CUSTOMER _X Y Z Company_   NO. _6199_

DATE _6/17/68_   ESTIMATOR _W.Q.K._   SHEET _1_ OF _1_

1	MATL. PREP.	✔	6	MACHINE	✔	11	SAND SURFACE ✔ 16 HANDLING
2	LAYUP	✔	7	ROUGH	✔	12	PRIME 17 UNASSIGNED TIME
3	INSPECTION	✔	8	BOND	✔	13	PAINT 18 SUPERVISION ✔
4	TRIM	✔	9	MECH. ASS'Y	✔	14	NEOPRENE COAT 19
5	HOLES		10	SALVAGE		15	PACKAGE 20

CODE	SEQUENCE OF OPERATIONS	NO. MEN	TIME, MIN.	MAN MIN.
1	Cut set of patterns	1	8	8
1	make 100 x 84 in. bag	1	5	5
2	mix resin	1	3	3
2	clean mold	1	4	4
2	spray mold release	1	3	3
2	spray gelcoat	1	6	6
2	layup part	1	14	14
2	bag	1	7	7
2	remove part	1	5	5
3	inspect	1	2	2
4	trim part	1	20	20
7	sandblast outside surfaces	1	8	8
11	sand out inside bag wrinkles	1	10	10
3	final inspection	1	3	3
18	supervision	1	12	12

AVERAGE PRODUCTION RATE   TOTAL LABOR, MAN MINUTES _110_

1 man makes 6 pcs. per day   MAN HOURS _1.84_

plus mat preparation,

inspection and supervision.

Fig. 5–32.   Labor estimate sheet (2).

COST ESTIMATING
ENGINEERING AND TOOLING

CUSTOMER _X Y Z Company_   NO. _6199_

DATE _6/17/68_   ESTIMATOR _W.Q.K._   SHEET _1_ OF _1_

1 ENGINEERING		5 MOLDS	✔	9 DRILL JIG		13 SETUP	✔
2 DESIGN		6 SCREEN		10 TRIM JIG	✔	14 TESTING	
3 DRAWINGS		7 CUTTING TEMP		11 MACH FIXT		15 HANDBOOKS	
4 PATTERNS	✔	8 INSP. FIXT		12 ASSEMBLY FIXT		16	

CODE	DESCRIPTION	BUY	MAKE	DELIV WEEKS	MAN HOURS	RATE	LABOR COST	MAT'L COST	AMORTIZED TOOL COST	NON-AMORTIZED TOOL COST
4	male wood pattern		✔	1	80	4.00	320.00	55.00		375.00
	(inside)									
5	female epoxy laminate		✔	1	70	4.00	280.00	157.00		437.00
	bag mold with									
	hold down ring									
7	mat cutting temp.		✔	1/2	4	4.00	16.00	2.00		18.00
10	trim jig (for rad. saw)		✔		12	4.00	48.00	6.00		54.00
13	tool setup		✔		2	4.00	8.00		8.00	

ESTIMATED DELIVERY   _2½ weeks_   TOTAL $8.00 $884.00

Fig. 5–33.   Engineering and tooling costs for plastic tank (2).

COST ESTIMATING SUMMARY SHEET

CUSTOMER _XYZ Company_   NO. _6199_

DATE _6/17/68_   ESTIMATOR _W.Q.K._   SHEET _1_ OF _1_

AREA, SQ. IN.	3141	MATERIAL ALLOWANCE	$ 12.21
AREA, SQ. FT.	21.8	LABOR ALLOWANCE, MAN MIN.	110
THICKNESS, IN.	0.100 ± 0.015	TOOL & ENGINEERING ALLOWANCE	
DENSITY, LB CU. IN.	0.0551	AMORTIZED	$ 8.00
WEIGHT OF LAMINATE, LBS.	17.3	NON-AMORTIZED	$ 884.00
WEIGHT OF FITTINGS, LBS.	–	OVERHEAD FACTOR	215
TOTAL WEIGHT, LBS.	17.3	G & A FACTOR	1.07
TRIM PERIMETER, IN.	106	PROFIT FACTOR	1.11
\ SQ. IN., PER IN.	29.6	LABOR RATE, PER MIN.	$ 0.03
WR	1.14		
WL	1.148		

TOOL DELIVERY _2½ weeks from acceptance of order_

PART DELIVERY _25 per week starting 4 weeks from acceptance of order_

QUANTITY	REJECTION ALLOWANCE	MATERIAL SELLING PRICE (1)	LABOR UNITS	LABOR UNITS QUANTITY	LABOR SELLING PRICE (2)	ENGRG. & TOOL SELLING PRICE (3)	TOTAL SELLING PRICE (4)
100	1.03	$14.95	2.04	2.04	$ 17.22	$ 0.09	$32.26

(1) MATERIAL SELLING PRICE EQUALS MATERIAL ALLOWANCE TIMES REJECTION FACTOR TIMES G & A FACTOR TIMES PROFIT FACTOR.

(2) LABOR SELLING PRICE EQUALS LABOR ALLOWANCE TIMES LABOR RATE TIMES LABOR UNITS DIVIDED BY QUANTITY TIMES OVERHEAD FACTOR TIMES G & A FACTOR TIMES PROFIT FACTOR.

(3) FOR AMORTIZED ENGINEERING AND TOOLING EXPENSE ONLY: ENGINEERING AND TOOLING ALLOWANCE DIVIDED BY QUANTITY TIMES G & A FACTOR TIMES PROFIT FACTOR.

(4) TOTAL SELLING PRICE - 1 + 2 + 3.

(5) FOR ENGINEERING AND TOOLING NOT AMORTIZED: ENGINEERING AND TOOLING ALLOWANCE TIMES G & A FACTOR $ 1050⁰⁰ TIMES PROFIT FACTOR.

ESTIMATED COST FOR 100 PIECES = 100 + $32.26 =   3226⁰⁰

MAXIMUM TOTAL ESTIMATED COST   4276⁰⁰

APPROVED

ESTIMATING _C.N. 6-12-68_   PLANNING _O.R. 6-13-68_

Fig. 5-34. Summary sheet for costs of plastic tank of Fig. 5-29 (2).

selection of equipment, abrasives, lubricants, and ratio of volume of work: volume of batch and barrel.

To arrive at accurate part or product manufacturing costs, the estimator must determine the cost of any necessary surface finishing operations. Determining the cost of such operations requires accurate data concerning abrasives, compounds, loading and unloading time, and machine time.

Equipment

In making his estimate, the estimator should compare the relative efficiency of the various machines available in his plant. Accurate cost records on previous operations assist in making sound estimating decisions.

TUMBLING BARREL FINISHING COST CALCULATOR

	COSTS $ PER YEAR	COSTS $ PER HOUR	COSTS $ PER PIECE
FLOOR SPACE COSTS			
ADD ALL THESE COSTS PER YEAR FOR A DEPARTMENT FLOOR AREA IN WHICH MACHINE IS LOCATED:			
RENT $____			
HEAT ____			
LIGHT ____			
INSURANCE ____			
GEN. OVERHEAD ____			
SUPERVISION ____			
TOTAL (A) $____ X ____SQ. FT. NEEDED FOR OPERATION / ____SQ. FT. IN DEPARTMENT	$____		
DEPRECIATION AND MAINTENANCE			
TOTAL COST OF TUMBLING BARREL & ACCESSORIES . . . $____			
DIVIDED BY YEARS OF DEPRECIATION ALLOWED . . . . ____YRS =	$____		
MAINTENANCE COST PER YEAR	$____		
TOTAL (B) . . .	$____		
MACHINE HOURLY COST RATE			
TOTAL (B) ABOVE $____ ÷ ____ HOURS USED PER YEAR		$____	
POWER COSTS PER HOUR (1, 2, 3, 5, 10HP USE $.02 PER HP PER HOUR)		$____	
MEDIA COST PER LOAD			
____LBS. PER LOAD X $____PER LB. X____ % ATTRITION RATE . . . . (SEE CHART ON OTHER SIDE)		$____	
TOTAL (C)		$____	
COST PER LOAD			
COMPOUND USED PER HOUR____LBS. X $____PER LB.			$____
TOTAL (C) HOURLY RATE $____ X ____HOURS PER LOAD			$____
DIRECT LABOR RATE (+ FRINGES) $____ PER HOUR X____HOURS PER LOAD . . . . (WORKER NEEDED FOR ONLY PART OF LOAD CYCLE TIME)			$____
TOTAL (D)			$____

TOTAL (D) $____ ÷ ____PIECES PER LOAD = $ ____ PER PIECE

Fig. 5–35. Cost estimate form for parts finished through the use of tumbling barrel equipment (3).

Finishing Costs per Piece

Shop performance is the most important factor in determining finishing costs per unit. Accurate shop performance costs require a close study of labor rates and times, material costs, and overall processing operations.

Estimating Procedures

Several methods can be used to calculate the cost of finishing parts through the use of tumbling or vibratory equipment. Fig. 5–35 shows the estimating form one company uses to estimate finishing costs with tumbling barrel equipment. As indicated by the estimating form (Tumbling Barrel Finishing Cost Calculator) shown in Fig. 5–35, the costs incurred in barrel finishing are: (1) floor space, (2) depreciation and maintenance, (3) power, (4) media (abrasive), and (5) labor.

The assignment of these cost elements to an individual piece part requires the following five-step procedure:

1) Determine floor space costs. The costs of rent, heat, light, insurance, general overhead, and supervision for the department using the barrel finishing equipment are totaled and multiplied by the ratio of space needed for finishing equipment : space in the entire department.

2) Compute annual depreciation and maintenance costs. This figure is added to "Total A, Floor Space Costs," to give "Total B."

3) Calculate the hourly machine rate. "Total B" (the sum of floor space costs and depreciation and maintenance) is divided by annual hours of operation. To this figure is added power costs per hour, and media cost per load. Determining media cost per load involves the use of attrition rates for finishing media. Attrition data are shown in Table V-24. The total of these items ("Total C") results in an hourly cost for operating the machine.

Table V-24. Media Attrition Data for a Tumbling Barrel Finishing Machine* (3).

Slide Systems (Surface FPM)	Random Shape, Alum. Oxide ($\frac{1}{2}'' \times \frac{9}{16}''$)	Preformed Cylinder, Silicon Carbide ($\frac{3}{8}''$ diam. × $\frac{3}{4}''$ long)	Preformed Triangle Alum. Oxide ($\frac{5}{8}'' \times \frac{3}{16}''$)	Preformed 45° Angle Cut Cylinder Alum. Oxide ($\frac{1}{2}''$ diam. × $\frac{7}{8}''$ long)	Ceramic Cones	Random Shape, Red Granite Size 2
25	.08	.07	.05	.02	.03	.02
50	.19	.15	.11	.05	.06	.04
75	.32	.24	.18	.08	.10	.06
100	.45	.34	.26	.11	.15	.08
125	.59	.45	.34	.15	.19	.10
150	.74	.56	.42	.18	.24	.12
175	.87	.66	.50	.21	.28	.14
200	1.02	.77	.59	.25	.33	.16
225	1.15	.87	.67	.28	.37	.18
250	1.29	.98	.75	.31	.42	.20

*Per cent of media consumed per hour.

4) Determine cost per load. The cost of the compound used per hour is first computed, and then the machine rate per load is determined. The labor cost per load is determined, and the three cost items per load are totaled to give cost per load ("Total D").

5) Determine finishing costs per piece. Cost per load ("Total D") is divided by pieces per load.

REFERENCES

1. Naujoks, Waldemar, and D. C. Fabel, *Forging Handbook* (The American Society for Metals, Cleveland, Ohio, 1939).
2. Sheffler, F. W., "Estimating for Reinforced Plastics," *Modern Plastics* (May, 1957).
3. J. F. Rampe, Rampe Manufacturing Company, Cleveland, Ohio.

BIBLIOGRAPHY

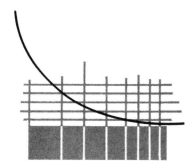

Ahlemeyer, C. H. "Economic Justification," *ASTME Paper No. MM63-648* (1963).

Bangasser, R., and J. Tonini. "Nomography—An Estimating Tool," *ASTME Paper No. MM66-144* (1966).

Benkelman, W. D. "A Re-Evaluation of Production Processes for Plastic Parts," *ASTME Paper No. MM66-140* (1966).

Beutel, M. L. "New Code Simplifies, Expands Cost Estimating by Computer," *Engineering* (May 14, 1964), 30–2.

Brescka, R. S. "Tool Room Cost Control," *ASTME Paper No. MS65-714* (1965).

Brown, C. F. "ABC's of Weld Estimating," *Welding Engineer* (October, 1958).

Camps-Camping, F. M. "Cutting Costs with Value Analysis," *ASTME Paper No. MS65AP039* (1965).

Cocker, R. "Cost Comparison Determines Machining Method," *ASTME Paper No. MM67T094* (1967).

"Computerized Cost Estimating." *Tool and Manufacturing Engineer, 57*, No. 5 (November, 1966), 13.

Conn, H. "Economic Justification of Equipment for Short Runs," *ASTME Paper No. MM66-575* (1966).

Corker, R. "Machining Cost Comparison—Milling, Grinding, Planning," *ASTME Paper No. MM66-173* (1966).

Doyle, L. E. "Cost Estimating, How To Minimize the Dangers of Chance," *The Tool Engineer, 42*, No. 7 (June 15, 1959).

Dudick, T. S. *Cost Controls for Industry* (Englewood Cliffs, New Jersey: Prentice-Hall, Inc., 1962).

"Estimating with a Computer," *American Machinist, 110* (November 21, 1966), 115–17.

Fleming, R. A. "Facilities Planning," *ASTME Paper No. MS66-907* (1966).

"For Your Operating Management—Unprecedented Control over Key Costs!" *Modern Materials Handling, 22* (October, 1967), 43–50.

Gallagher, P. F. *Project Estimating by Engineering Methods* (New York: Hayden Book Companies, 1964).

Gillespie, C. *Standard and Direct Costing.* Revised Edition of *Accounting Procedures for Standard Costs* (Englewood Cliffs, New Jersey: Prentice-Hall, Inc., 1962).

Gould, A. F. "Operations Research Case Studies," *ASTME Paper No. MM62-103* (1962).

Graham, C. F. *Work Measurement and Cost Control* (New York: Pergamon Press, Inc., 1965).

Hammerton, J. C. "Cost Determinants in Designing Production Control Systems," *Automation, 14* (November, 1967), 92–7.

Harig, H. "Accurate Pricing of Dies," *ASTME Paper No. MS65-121* (1965).

Henrici, S. B. *Standard Costs for Manufacturing* (3d ed.; New York: McGraw-Hill Book Company, Inc., 1960).

Hill, L. S. "Towards An Improved Basis of Estimating and Controlling R & D Tasks," *Journal of Industrial Engineering, 18* (August, 1967), 482–8.

Huff, A. "Optimum Speeds for Automatics," *American Machinist/Metalworking Manufacturing, 107* (April 15, 1963), 93–6.

Jacobs, H. J. "Estimating Short Run," *ASTME Paper No. MM65-606* (1965).

Kellee, L. W. "The Use of a Computer in Cost Estimating," *ASTME Paper No. MM67-680* (1967).

McNeill, T. F., and D. S. Clark. *Cost Estimating and Contract Pricing* (New York: American Elsevier Publishing Co., Inc., 1966).

Naujoks, Waldemar, and D. C. Fabel. *Forging Handbook* (Cleveland, Ohio: The American Society for Metals, 1939).

Nichols, W. T. "Capital Cost Estimating," *Industrial and Engineering Chemistry* (October, 1951).

Petruschell, R. L. "Project Cost Estimating," *Royal Aeronautical Society Journal, 71* (November, 1967), 737–44.

Quartarone, R. "The Economic Considerations of Numerical Control," *ASTME Paper No. MM63-666* (1963).

Reichert, D. I. "Estimating for Short Run Production of Electronic Systems," *ASTME Paper No. MM66-701* (1966).

Sluhan, C. A. "How Cutting and Grinding Fluids Effect Value Analysis — Manufacturing Cost Reduction," *ASTME Paper No. MM63-565* (1963).

Wallis, B. J. "The Economies of Transfer Dies," *ASTME Paper No. MF66-133* (1966).

Williams, R., Jr. "Why Cost Estimates Go Astray," *Chemical Engineering Progress, 60* (April, 1964), 15–18.

Winn, L. J. "Selecting of Optimum Proposals from Various Value Engineering Alternatives," *ASTME Paper No. MM64-597* (1964).

Zeyher, L. R. *Cost Reduction in the Plant* (Englewood Cliffs, New Jersey: Prentice-Hall, Inc., 1966).

INDEX

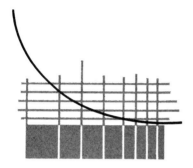

A

Accounting department, 1, 6, 20, 36, 47
  compared with estimating, 6
  estimating responsibilities, 5, 10, 19, 84, 93, 111
Accuracy, 16, 26, 28, 36, 39, 96
Actual costs (*See* Standard costs)
Alternative designs, 7, 23
Alternative specifications, 64
Aluminum-base alloys, 104
Aluminum die casting estimate (*See* Cam drive bracket)
Aluminum forging, 58-64
Amortized engineering and tooling expense (*formula*), 99
Annealing, 84, 91
Annodizing, 92
Arc time, 91, 98
Audio receivers, 17
Automatic feed, 62
Automatic feeding equipment, 110 (*See also* Stamped parts)
Automatic fusion welding, 91
Automatic screw machine (*See* Screw machine)
Automatic welding, 90

B

Balance sheet, 5
"Ballpark" figures, 14
Bar end loss, 104
Barrel finishing, 118-122 (*See also* Tumbling and vibratory finishing)

Bending, 101
Bevel welds, 97
Biased errors, 31
Bid requests, 7, 20
  overestimating, 30
Bill of materials, 36, 37
"Blank nesting," 111
Blow plates, 82
Blueprints, 34
Box tool, 75
Box-turning, 65
Bracket assembly, 51-58
  assembly drawing (*figure*), 52
Breakdowns, 28
Break-even point, 24
Brushing, 92
Building requirements, 36, 37, 39
Burden (*See* Factory burden)
Burn-off rate, 91

C

Calculation sheet (*figure*), 117
Cam drive bracket, 51-58
  assembly drawing (*figure*), 52
Cam, indexing data (*table*), 76
Carbide cutters, 115
Cash flow statement, 5
Castings (*See* Sand castings)
Certification, 93, 94
Circular cutoff tools (*table*), 72
Cleaning, 47, 106, 118
  cleaning costs, 83, 89
  forged parts, 106
Clearances (*table*), 77

Clerical personnel, 17 (*See also* Personnel)
Closed-die forgings, 100, 104
Cold-trimmed forgings, 103
Communication, 19
Company policy, 19, 20
Company practice, 49
Comparison method, 26, 27
  estimating data, 26
  used with conference method, 27
Competition, 12
Competitive analysis, 6, 13, 14
Completed estimate (*See also* Cost
    estimating)
  adjustments, 20
  approval, 20, 21
  example (*figure*), 43
  forwarded to accounting, 93
  reference copy, 62
  reviewed, 21
  routing, 19
Composite pattern plates, 82
Compound costs, 122
Compression molds, 115
Computerized estimating, 15-18
  advantages, 16-17
  estimate preparation, 16
  feasibility, 16
  problems, 18
  training, 18
Conference method, 26
  advantages, 26
  disadvantages, 26
  participation, 26
  used with comparison method, 27
  used within estimating department, 26
Construction cost estimate, 2
Contingency factors, 29
Contracts, 12, 20, 90, 94, 99
Controlling the estimate, 28-33
Conversion burden and labor, 113
Copper-base alloys, 104
Cores, 82-83, 88
  core assembly (*figure*), 88
  core costs, 83
  cost estimate (*figure*), 88
  definition, 83
  tooling, 82, 83
Corporate strategy, 10
Corrective welding, 92
Correspondence, 34
Cost (*See* Manufacturing costs)
Cost accounting, 4, 5, 33
Cost analysis request form (*figure*), 24
Cost centers, 47, 49
Cost control, 1, 5-6
  cost requests, 21-22, 24
  responsibilities, 5
Cost data, 6, 19, 35, 36, 43, 45

Cost deviations, 29
Cost environment, 31-32
Cost estimating
  accuracy requirements, 26, 28, 36, 37
  action copy, 62, 64
  classes of estimates, 2, 7, 37
  compared with other functions, 4
  components, 1
  courtesy estimates, 20
  definition, 1
  estimating systems, 10-12
  examples, 53, 60, 88, 98
  initiation, 19
  lead time, 11
  manufacturing cost structure, 6
  preparation costs, 22, 29
  processing steps (*figure*), 21
  product development cycle, 12
  purchased parts, 64
  relation to other functions, 4
  review, 21
  role, 2
  routing sheet (*figure*), 89
  selling price, 74
  terminology, 1-2
  time requirements, 36
  uses, 7
    aid to management, 6
    competitive analysis, 6
    customer orders, 22
    design selection, 23
    evaluation and comparison, 7
    information for management, 5
    make-or-buy decisions, 4
    purpose, 6-7
    temporary standards, 7
    vendor analysis, 6
Cost indexes, 28, 33
Cost reduction, 5-6, 19, 22, 24
Cost requests, 19, 22, 23, 24, 51, 116
  analysis, 35, 36
  design specifications, 36
  due date, 36
  estimate request form (*figure*), 22
  examination, 19
  for water outlet, 84
  initiated, 21-26
  missing information, 36
  routings, 19
  screening, 19-20, 22
  stamped parts, 109
Courtesy estimates, 20
Cover outlet, 93-99
Cross-slide forming, 65
Cross-slide tool, 65
Curing, 47
Customers, 20, 22, 34, 64, 116
  customer orders, 22

Customers (continued)
  customer quotation, 64
  customer specifications, 77
Cut diameters, 62
Cutoff losses, 94
Cutoff tools
  angle and thicknesses (table), 72
  spindle revolutions, 75
Cutting operations
  costs, 99
  cutter approach (formula), 57
  overlapping operations, 67
  production economy, 67
  speeds and feeds
    for steels (table), 68-69
    side milling cutter (formula), 55-56
  tools included as burden, 63
  wear rates, 62
Cut waste, 104
Cycle times, 114

D

Data, 20
  data file, 13, 36, 43, 46
  handbooks, 62
  storage by computer, 16, 18
Deburring, 118
Defective material, 28
Delivery specifications, 64
Density of metals (table), 102
Departmental organization, 9, 12-13
Deposited metal, 96-97, 98
Depreciation, 121, 122
Design
  costs, 23, 32
  design engineering department, 7
  estimate request, 23
  evaluation of design proposals, 7
  specifications, 36
Detail drawings, 36
Detailed analysis method, 27, 32 (See also
  Final cost estimate)
Dies
  costs, 104-105, 108, 112
  die blocks, 104
  die life, 104, 106
  flat dies, 100
  impression dies, 100
  resinkings, 104-105
Dipping, 47
Direct costs, 2, 3 (See also Manufacturing
  costs)
  labor, 47, 63, 91-92
  material, 41-44, 104
Drawings, 20, 21, 22
Drop forgings (See Closed-die forgings)

Durable tooling, 3, 26, 45
  burden item, 3, 45, 92
  costs, 36

E

Economic trends, 32
Electrodes, 90
Engineering changes, 13, 27, 34
Equipment, 7, 27
  capacity, 19
  certification, 92
  costs, 49, 120
  installation, 39
  performance factor, 28
  requirements, 36
  slitting and shearing, 42
Errors, 16, 28, 32, 38
Estimate analysis, 36-37
Estimate control devices, 28
Estimate deviations, 28, 30, 31
Estimate examples
  die castings, 51-58
  forged parts, 100-109
  machining costs, 58-64
  plastic parts, 116-118
  sand casting, 77-89
  screw machine forming, 64-77
  stamped parts, 109-113
  tumbling and vibratory finishing costs,
    118-120, 122
  welded items, 93-99
Estimate request, 21, 77 (See also Cost
  requests)
Estimating (See Cost estimating)
Estimating conference, 7, 26
Estimating data (See also Standard data,
  Data), 10, 16, 17, 22, 26, 34, 35, 60
Estimating department, 1, 19, 116
  administrative supervisor, 10
  conference method, 26
  development time, 10
  in the manufacturing organization,
    10-12
  long-range planning, 2
  management, 9, 10, 13
  operating costs, 12
  organization, 12-13
  relationship to accounting department,
    6
Estimating forms, 20, 33, 43, 51, 73
  completion of, 62
  distribution, 60
  examples, 8, 22, 23, 24, 25, 27, 42, 43,
    53, 55, 60, 61, 66, 87, 89, 95, 96,
    97, 98, 101, 107, 110, 116, 117,
    118, 119, 120, 121
Estimating procedures, 14, 26-27, 35-50

Estimating Procedures *(continued)*
  for specific products, 94, 100–101, 110
    121
  large operations, 35
  methods, 26–27
  quality requirements, 92
  simplification, 28
  small operations, 35
  speeded by data files, 36
Executive salaries, 3
Expensive products, 28

F

Facilities, 2, 3, 21, 39, 40
Factory burden, 2, 3, 10, 26, 48–49, 63,
    73–74
  burden items, 3
  durable tooling, 92
  estimator's responsibility, 2
  factory housekeeping, 3
  nonproductive costs, 63
  rate, 64
  standard cutting tools, 63
  starting costs, 63
  tooling, 93
  utilities, 93
Fatigue, 73, 100
Faulty tooling, 30
Feed rate, 75, 115
Filler metal, 96
Final cost estimate, 7, 8, 42
Finished castings, 78
Finishing, 36, 92, 121
Finish machining, 92
Fitup, 91, 92
Fixed costs, 3, 32 *(See also* Manufacturing
    costs)
Fixtures, 92, 96
Flash, 103, 107
Flasks, 82
Flat-die forgings, 100
Flat-dies, 100, 104
Floor space, 121, 122
Forecasting, 32
Forged parts, 100–109
Forging equipment costs, 105
Forging machine, 100
Form tool, 75
Foundry burden, 84
Foundry tooling, 82
Full-height cam *(table)*, 76
Furnace charge, 78

G

Gages, 27, 82
Gap width, 92

Gear blank, 106–109
Gelcoat, 117
General and administrative costs, 2, 3, 64,
    65, 72, 89, 106, 113
Grain flow position, 100
Gray iron castings, 79, 84, 102–103, 107
"Guesstimate," 30

H

Hammered forgings, 100
Handbooks, 46, 62, 78
Hand cleaning, 83
Hand forgings, 100
Hardening, 91
Heat treating, 38, 41, 77, 84, 91, 106
High estimates, 30
High-production manufacturing methods,
    34
High-speed steel tools, 115
Hill-to-valley relationship, 75
Historical data, 6, 26, 27, 30, 33, 36, 45
Hole reaming, 52
Honing, 118
Hot-trimmed forgings, 103
  flash thickness and width *(table)*, 103

I

Impression dies, 100, 104
Indexing data *(table)*, 76
Indirect costs, 2, 3, *(See also*
    Manufacturing costs)
Indirect material, 3, 115 *(See also*
    Manufacturing costs)
Industrial engineers, 17
In-house jobs, 14
In-house parts, 41
In-house tooling, 44–45, 92, 96, 99
Injection molds, 115
In-process gaging, 62
Inspection, 14, 92–93, 113
  costs, 84, 99
  destructive, 93
  equipment, 27, 36
  included as factory burden, 63
  level of inspection, 94
  nondestructive, 93
  of stamped parts, 110
  operations, 77
  procedures, 36
  visual, 93
  water outlet casting, 89
  welder inspection, 92
Integrated estimating system, 11–12
Internal cuts, 62
Iron castings, 84

J

Job grades, 62
Job loyalty, 15
Job shops, 12, 13, 35, 46
Joining weld, 97

L

Labor, 2, 7, 25
  costs, 2, 26, 33, 46, 47, 63 *(table)*, 73, 76
    assignment, 53
    changes, 32
    examples, 81, 96, 99, 106, 114, 118, 121, 122
    high estimates, 30
    labor estimate sheet *(figure)*, 119
    labor summary sheet *(figure)*, 97
  fringe benefits
    group insurance, 47
    retirement plans, 47
    social security, 47
  labor efficiency rate, 73
  labor operations, 64
  labor problems, 30
  labor rates, 10, 40 *(table)*, 63, 73
  labor selling price *(formula)*, 99
  labor times, 94
  skills, 19
Laminate, 117
Laying out cams *(table)*, 70
Layouts, 7, 36, 66, 112
Lead time, 6, 11, 21, 39, 41, 46
Learning curve, 17
Lever, 109-113 *(See also* Stamped parts)
Lighting, 122
Limit tables, 17
"Look-up tables," 18
Lost metal *(formula)*, 87
Low estimates, 30

M

Machinability for steels *(table)*, 68-69
Machine certification, 94
Machine forgings *(See* Closed-die forgings)
Machining losses, 94
Machining operations, 76, 105-106
  machining costs
    aluminum forging, 58-63
    barrel finishing, 122
    gear blank, 108
    outlet cover, 99
    pin *(figure)*, 66
    sand castings, 83-84
  machining time, 52, 57, 62, 75
    operation time, 46
    screw machine, 64
    steps to determine machine time, 66

Machining Operations *(continued)*
    straddle milling, 55
    performance factors, 28
    spindle revolutions *(table)*, 73, 75
    stamped parts, 112
Maintenance, 46, 121, 122
Make-or-buy decisions, 1, 4, 6, 7, 8, 25, 26, 38
Malleable iron castings, 80
Management, 4, 9, 11, 13, 17, 93
Manufacturing cost analyst, 10
Manufacturing costs, 1, 2-6
  actual, 2, 3
  cost projections, 17
  customer relations, 20
  direct, 2, 101-104
  estimated by computer, 15-18
  fixed, 3, 32, 108
  for alternative methods, 24
  for outlet cover, 98
  general and administrative costs, 106
  increases, 32
  indirect, 2-3
  machine costs, 108
  manufacturing cost structure, 2-6
  predetermined, 22
  reduction of, 5
  responsibility, 4-6
  sand casting, 89
  segregation of, 20
  standard, 2, 3
  step-variable, 3
  summarized, 21
  tooling, 35, 104, 108
  variable, 3
Manufacturing engineering department, 10, 19, 21-22, 24, 39
Manufacturing operations, 38-39
  economy, 7
  feasibility, 22
  manufacturing plan, 6, 21, 24
  methods, 7, 24, 34, 39
  procedures, 14, 34
  processes, 14, 22, 28
  process plan, 35, 36, 51
  routings, 36
  techniques, 14, 24-25, 39
  times, 21, 36, 45-47
Marketing department, 20
  estimate request *(figure)*, 22
  initiates cost requests, 21-22
  initiates estimates, 19
  selling price, 49
Material, 2, 7, 14, 23, 25
  allowances for forged parts, 101
  characteristics, 74, 77
  cost data, 43
  defective, 28

Material (continued)
  for water outlet, 84
  material requirement examples, 94,
    106, 111, 117
  price breaks, 42
  purchased, 25
  quantity purchases, 42, 74
  reclamation allowance, 42
  scrap, 42
  selection, 12
  shortages, 30
  size, 74
  wastage, 3
Material allowance/finished weight ratio,
  101-102
Material costs, 23, 26, 28, 36, 65, 74, 76,
  78-81, 104
  changes, 32
  examples, 85, 94, 107, 110, 111, 115,
    118
  reduction of, 42
  stamped parts, 110, 111
  weight calculations (table), 108
Material handling, 6, 21, 39
Material lossage, 94
Material summary sheet (figure), 95
Media, 121, 122
Metal stampings (See Stamped parts)
Methods analysis, 14
Methods engineering department, 12, 19,
  20
Mirror finishing, 118
Molds, 82, 85, 89, 115
Multiple cavity dies, 114

N

Net weight, 102, 107
Nonamortized engineering and tooling
  (formula), 99
Nonproductive operations, 63, 66, 72, 76

O

On-the-job training, 9, 12
Open-die forgings, 100, 102-103, 104
Operations, 92
  operation lineup, 39
  operation sequence, 72, 75
  operation sheets, 20
  reports, 17
  time, 46
Optical scanning terminals, 17
Organization, 9-19
  departmental, 12-13
  influencing factors, 9
  organizational ambiguity, 12
  typical organization (figure), 11

Overhead, 31, 32, 33, 81, 83, 89, 114, 122
Overruns, 78

P

Packaging, 93, 115
Painting, 47, 92, 99
Part analysis, 37-38, 65, 114
Performance standards, 8, 14
Perishable tooling, 35, 45, 92
Personnel, 73
  diversified background, 15
  experience, 13, 14, 15
  formal education, 15
  personnel policies
    acquisition, 14
    efficient use of personnel, 17
    job loyalty, 15
    professional literature, 15
    provide working tools, 15
    staffing level, 12
    working environment, 15
  programming knowledge, 17
  qualifications, 9, 13-14
  training, 15, 18
Planimeter, 78
Plant engineering department, 19, 20, 21
Plant layout, 6, 17, 93
Plastic fabricating processes, 113, 114
Plastic parts, 113-118
  burden, 114
  cycle time, 114
  direct material, 115
  estimate summary, 118
  indirect material, 115
  labor costs, 114
  overhead, 114
  tooling costs, 114, 115, 118
  weight calculations, 115
Poured metal, 81, 86, 87
Predetermined costs (See Standard costs)
Preliminary estimates, 7-8
  estimating form (figure), 8
  preparation, 8
  purpose, 7
Preliminary manufacturing plan, 38-39
Preliminary process plan, 39
Press forgings (See Closed-die forgings)
Process engineering department, 7, 59
Process industries, 12, 21
Process planning, 6, 15, 38
  alternative manufacturing plans, 24
  plastic parts, 114
  process sheets, 51
  requirements, 65
  stamped parts, 109-110
  welding, 90
Process planning department, 12, 19, 20,
  21, 26

Process treatment times, 17, 47
Product
 complexity, 9
 configuration, 12
 costs, 6
 deficiencies, 23
 definition, 1
 design, 6, 17, 23, 24–25, 34
 development cycle, 12
 development decisions, 5
 drawings, 7
 product flow pattern, 39
 product mix, 5, 12, 13, 19, 20, 35
 product specifications, 20
 quality, 6
Product engineering department, 19,
 21–22, 23
Product estimate, 1
Production quantity, 3, 12, 22, 24, 32, 36,
 82, 100
Production scheduling, 8, 19, 27
Professionalism, 9, 15, 18
Profit, 10, 14, 20, 49, 64, 74, 89, 106, 109
Profit and loss statement, 5
Project cost estimate, 2
Prototypes, 22
Purchase cost analyst, 10
Purchased items
 components, 90
 materials, 21
 parts, 21, 25, 34, 41, 51–52, 59, 64, 94
 purchasing burden, 90
 tooling, 44, 92
Purchasing department, 4, 12, 19, 20, 24,
 26, 59

Q

Qualifications for estimators, 9, 13–14
Quality control (See Inspection, Testing)
Quality requirements, 14, 90, 92
"Quantity extra," 107
Quenching, 84

R

Raw material, 27, 41–42
Raw stock (See also Stock, Vendors), 65, 94
Reclamation allowance, 42
Records, 35, 93
Reliability requirements, 14
Remelted metal, 81, 86
Research and development, 3
Resinking, 104–105, 108
Resistance welding, 91, 94
Return on investment, 7, 8
Revolutions (See Spindle speeds)
Routing sheet, 87, 89

Routing sheet (continued)
 cost estimate (figure), 89
 for water outlet casting (figure), 87

S

Sales department, 2, 10, 20, 22, 64, 106,
 109, 116
Salvage, 27, 110, 113
Sand castings, 77–89
 foundry burden, 84
 inspection costs, 84
 labor cost, 81
 machining costs, 83–84
 material cost, 78–81
 overhead, 81, 83, 89
 profit, 89
 shipping costs, 84
 tooling costs, 83
Scale, 103, 107
Scrap allowance, 5, 27, 42, 90, 94, 106,
 109, 111, 112, 115
Screw machine, 64–77
 factory burden, 73
 general and administrative costs, 74
 labor costs, 73, 76
 manufacturing costs, 64, 65, 66, 73, 74,
 76
 material costs, 74, 76
 setup time, 73
Selling price, 2, 49–50, 64, 74, 99, 111
Setups, 26, 63, 73, 91, 96, 106
Shape weight, 102, 106–107
Shearing, 106, 110, 113
Shipping costs, 84, 93, 113
Shop yield (formula), 78
Side milling cutter, 55–56, 57
Single-spindle screw machine (See Screw
 machine)
Slag, 92
Smith forgings (See Open-die forgings)
Special equipment, 93
Special tools, 74, 92, 96, 104–105
Spindle revolutions, 70 (table), 73, 75
Spoilage rates, 26
Spot welding, 90, 98, 99
Staffing, 9–18
Stamped parts, 109–113
 conversion burden, 113
 conversion labor, 113
 die cost, 112
 estimating procedures, 110
 factory burden, 110, 112
 general and administrative costs, 113
 inspection, 110, 113
 labor costs, 110, 112
 material cost, 110, 111
 scrap allowance, 111, 112
 selling price, 111

Stamped parts *(continued)*
  total manufacturing cost, 113
Standard costs, 2, 3, 5, 22, 28, 29, 44 (*See
  also* Manufacturing costs)
Standard data, 33, 62, 63, 82
  data for spot welding *(table)*, 99
  estimator's chart *(table)*, 105
  for barrel finishing *(table)*, 122
  for drilling, reaming, and tapping
    aluminum *(table)*, 56
  for estimating handling and machine
    times, 52
  for fillet and bevel welds *(table)*, 97
  for forged parts, 101
  for milling machine *(table)*, 56
  for time study *(table)*, 54
  for turret lathe *(figure)*, 61
Step-variable costs, 3-4
Stock, 41, 65, 72, 76, 78, 91, 111
Straddle milling, 55, 57
Subassemblies, 1, 36
Subcontracting, 36, 38
Summary sheets, 20, 95, 96, 97, 98, 120
Supervision, 3, 122
Surface finishing, 41, 92, 94, 96
Surface oxidation, 103

T

Table blasting, 83
Templates, 92, 96
Testing, 27, 36, 93
Thermostat housing, 84
Threaded inserts, 115
Time-and-motion study, 9, 14, 15
  operation sheet, 53
  standard data *(table)*, 54
  time data, 17
Tolerances, 14, 23, 34, 36, 65
Tonghold, 104
Tooling
  breakage, 3
  clearance, 76, 77
  costs, 2, 6, 10, 21, 26, 35, 39, 44-45,
    49, 64, 65, 66-67, 83
  assignment of, 2, 45
  changes, 32
  data, 45
  die machining, 108
  die resinkings, 108
  durable tooling, 92
  for gear blank, 108
  for metal stamping dies, 45
  for outlet cover, 94, 96
  for plastic parts, 114, 115, 118
  for plastic tank *(figure)*, 119
  perishable tools, 92

Tooling *(continued)*
  prorated, 45
  design, 6, 10, 51
  inspection, 62
  layout, 65, 74
  tool design department, 26
  tool engineering department, 59
  tooling estimate, 31
  tools, 25, 72, 75, 83, 92, 93, 115
Total manufacturing cost (*See
  Manufacturing costs*)
Transportation, 64, 90
Tumbling and vibratory finishing, 118,
  120-122
Turning, 75, 115
Turret tool, 65

U

Upset forgings (*See* Closed-die forgings)
Utilities, 3, 39

V

Value added, 6, 7
Value analysis, 1, 5-6, 14, 21-22, 38
Value engineering, 1, 5-6, 22, 24
Variable costs, 3
Variable operations, 62
Variances (*See* Standard costs)
Vendors, 7, 17, 25, 94
  quotations, 7, 25
  vendor analysis, 6, 8, 25-26
Volume requirements, 20

W

Waste factors, 117-118
Water outlet casting, 84-89
Wear rates, 62
Wear resistance, 84
Weight calculations, 108, 115
Welding
  electrodes, 90-91
  factory burden, 93
  methods, 96
  operations, 90-92
  processes, 90, 91, 94
  quality control, 90-93
  tooling, 92, 96
  weld deposit, 91, 94
  weld design, 91
  welding summary sheet *(figure)*, 96
Wholesale distributors, 12
Work assignment, 13
Worker performance factors, 28
Work methods, 32

Other Books from SME . . .

Manufacturing Data Series

Adhesives in Modern Manufacturing

Cold Bending and Forming Tube and Other Sections

Cutting and Grinding Fluids: Selection and Application

Cutting Tool Material Selection

Design of Cutting Tools: Use of Metal Cutting Theory

Functional Gaging of Positionally Toleranced Parts

Functional Inspection Techniques

**Gundrilling, Trepanning, and
  Deep Hole Machining (Revised Edition)**

High-Velocity Forming of Metals (Revised Edition)

Machining the Space-Age Metals

Non-Traditional Machining Processes

Pneumatic Controls for Industrial Application

Premachining Planning and Tool Presetting

**Producibility/Machinability of Space-Age and
  Conventional Materials**

Tool Engineering: Organization and Operation

Numerical Control Series

Introduction to Numerical Control in Manufacturing

Manufacturing Management Series

Introduction to Manufacturing Management

For further information, write to:

Society of Manufacturing Engineers
Publication Sales Department
20501 Ford Road
Dearborn, Michigan 48128

Realistic Cost Estimating Workbook

This section of *Realistic Cost Estimating* has been added to the original volume as an aid to students interested in testing their knowledge of the text. Consisting of four units, each related to the first four sections of the book, it asks the reader to identify the terminology of cost estimating, to provide expository answers to specific questions, and to indicate choice for true-false and multiple choice questions.

As an additional aid to those studying this book informally, answers to all questions are provided at the end of the workbook. It is SME's hope that the addition of this workbook to the original text will do much to assist readers in understanding all aspects of the increasingly important field of cost estimating.

The Editors

THE ESTIMATING FUNCTION

This unit covers Chapter I, pp. 1–8, in the text. The basis for the material studied in this unit is found within the content of that chapter. The text defines and explains the terminology used by cost estimators. The technical/financial terms illustrate the association of manufacturing technical concepts with many related financial and product planning functions.

Internal communications, written or verbal, are the greatest problems confronting modern business organizations. Because data and information created by the cost estimator are used as the basis for major decisions by other related business functions, it is essential that the user have a clear understanding of the meaning of each term expressed by a cost estimator. Misuse of estimated data because of mistaken interpretations is a chronic problem faced by manufacturing managers. A cost estimator makes a significant contribution to his business operation by always emphasizing the precise use of the correct terminology.

Definitions. Briefly define the following:

1. Cost estimating. _____

_____

2. Product cost estimate. _____

_____

3. Piece part. _____

_____

4. Subassembly. _____

_____

5. Direct costs. _____

_____

6. Direct labor cost. _____

_____

7. Indirect costs. _____

_____

8. Indirect labor cost. _____

_____

9. Factory burden. _____

_____

10. General and administrative overhead. _____

_____

11. Standard costs. _____

_____

12. Actual costs. _____

_____

13. Fixed costs. _____

_____

14. Variable costs. _____

_____

15. Step-variable costs. _____

_____

16. Cost accounting. _____

_____

17. Cost control. _____

_____

18. Value analysis. _____

_____

19. Value engineering. _____

_____

20. Make-or-buy decision. _____

_____

21. Preliminary cost estimate. _____

_____

22. Final cost estimate. _____

_____

23. Matchbook cover specifications. _____

_____

24. Tooling cost estimate. _____

_____

25. Facilities cost estimate. _____

_____

Short Answer Questions

26. Who has the ultimate responsibility for all financial aspects of a manufacturing firm's operation? Name three aspects. _____

_____

_____

27. What value does estimating add to a product? _____

_____

_____

28. Which aspect of cost estimating makes it distinctive from other financial functions? _____

_____

_____

29. Why is the cost estimating function important to a manufacturing organization? _____

_____

_____

30. What two questions does an estimator ask before classifying a cost as direct or indirect? _____

_____

_____

31. When are upper and lower limits of production quantities assumed to exist? _____

_____

_____

32. Describe two types of management decisions for which good cost estimating may be used. _____

_____

_____

33. Describe the way product engineering departments use the data from preliminary estimates. _____

_____

_____

34. What does effective value engineering require? _____

_____

_____

35. What purpose does a final estimate serve? _____

_____

_____

True or False

36. Standard costs are developed by the cost estimator. True _____ False _____

37. General and administrative costs are not part of the manufacturing cost. True _____ False _____

38. Direct costs include special tools for the part. True _____ False _____

39. Relatively small costs may be classified as indirect costs. True _____ False _____

40. Cost estimating provides information to evaluate last year's decision. True _____ False _____

41. Management evaluates departmental efficiency on the basis of variances from the original estimates. True _____ False _____

42. Cost control, value analysis, and value engineering are three related concepts. True _____ False _____

43. The most precise type of estimate is always worth the work involved. True _____ False _____

44. Direct and indirect costs are the same as standard and actual costs. True _____ False _____

45. The term *part estimate* could also mean *cost estimate.* True _____ False _____

46. The product cost estimate is the total estimated cost to manufacture a product. True _____ False _____

47. Estimating is generally a staff function. True _____ False _____

48. Estimates used for product costing may be classified in only one way. True _____ False _____

49. A subassembly has two or more piece parts. True _____ False _____

50. A product can have more than one subassembly. True _____ False _____

51. Accounting departments determine actual costs. True _____ False _____

52. Estimators or accountants determine direct costs. True _____ False _____

53. All tooling expenses are direct costs. True _____ False _____

54. Estimators determine the general and administrative expense. True _____ False _____

55. Precise wording is most important in reporting cost data. True _____ False _____

Matching. Select the appropriate matching word or phrase and record the letter for the word or phrase in the space to the left of each of the following statements.

(a) Actual costs
(b) Indirect costs
(c) Make-or-buy
(d) Value engineering
(e) Preliminary estimate

(f) Variances
(g) Accounting
(h) Factory burden
(i) Step-variable costs
(j) Profitability

56. _____ Concerned with the recording, analysis, and interpretation of events that have already transpired.

57. _____ Decisions based on cost estimates of internal manufacturing or the feasibility of purchasing an item.

58. _____ Reports of differences between what a part should cost and what it actually costs to produce.

59. _____ Costs for heat and lighting in the plant, janitorial services, machine repair workmen, etc.

60. _____ The objective determined by means of detailed estimating.

61. _____ Costs which include factory burden and general and administrative overhead.

62. _____ The costs for the finished product.

63. _____ A quick estimate of the cost for a new design proposal.

64. _____ Costs that remain the same over a given number of production units, then move to new plateaus when the volume changes.

65. _____ A systematic method to optimize product profitability.

UNIT ▌▌

ORGANIZATION AND STAFFING FOR ESTIMATING

Understanding the topics covered in Chapter 2, pp. 9–18, is an important requirement for a cost estimator. Technical and financial knowledge are prerequisites for generating estimated cost values, and an understanding of the future use of such cost values is necessary for the estimator to be self-directing in his creative efforts. The realm in which a cost estimator works is constantly expanding, and therefore the estimator must grow within each facet of the business operation. When a cost estimator understands his role in the decision-making process he senses his strengths and weaknesses. Cost estimators must undergo a continuous learning process in order to remain qualified to handle the problems arising from new products, new materials, new methods, or changing organizations. Whether cost estimating in a particular business is an individual or departmental staff function, the intelligent use of each individual's skills motivates the development of all personnel and stimulate the continual growth of estimating expertise.

Definitions. Briefly define the following:

1. Delegated responsibilities. _____

_____

2. Product complexity. _____

_____

3. Time-and-motion study. _____

_____

4. Price breaks. _____

_____

5. Job shop. _____

_____

6. "Ballpark" figures. _____

_____

Short Answer Questions

7. What are two basic personal attributes an estimator should have?

_____

_____

8. What is the difference between a bid estimate and a cost estimate?

_____

_____

9. List four planning functions (departments) or major areas in a large business organization whose activities are closely related to the cost estimating department. _____

_____

_____

10. Name and describe the two types of cost estimating systems commonly used by industry. _____

_____

_____

11. Describe the qualifications which should be used to evaluate candidates for estimator positions. _____

_____

_____

12. Why should an estimator not be part of the sales department in most companies? _____

_____

_____

13. What advantages are gained by not having estimators who are specialized in their talents? _____

_____

_____

14. What advantages can be gained by the specialization of estimators?

_____

_____

15. What key factors justify the use of computers for estimating? Describe at least three. _____

_____

_____

16. How can computers be used in estimating costs? _____

_____

_____

True or False

17. A computer will alert the estimator to his mistakes in judgment by the use of limit tables. True _____ False _____

18. The importance and use of cost estimates will decrease with greater computerization. True _____ False _____

19. Studies indicate that a company cannot afford to be without competent estimators. True _____ False _____

20. Price breaks are unethical sales practices. True _____ False _____

21. Almost any individual can become a good estimator. True _____ False _____

22. Top management must decide what kind of estimating organization is best. True _____ False _____

23. Wholesale distributors can afford more time for developing estimates than a manufacturing company can. True _____ False _____

24. A nonspecialized department is more flexible. True _____ False _____

25. Bias is an occupational hazard in cost estimating. True _____ False _____

26. An estimator must also be a rhetorician. True _____ False _____

27. The chief advantage of computer data storage is accuracy. True _____ False _____

28. Estimators must make technical decisions before a computer can complete the estimating program. True _____ False _____

Matching. Select the appropriate matching phrase and record the letter for the phrase in the space to the left of each of the following statements.

(a) Delegated responsibilities
(b) Integrated estimating system
(c) Time-and-motion study
(d) Cost control
(e) Input terminals
(f) Lead time
(g) Basic time data
(h) Organizational ambiguities
(i) Departmentalized estimating
(j) "Little black book"

29. _____ Utilizes a department which makes the complete estimate.

30. _____ A primary determining factor in departmental organization.

31. _____ Particularly critical requirement for a cost-estimate study.

32. _____ One key factor for using computers in estimating work.

33. _____ Equipment that should be readily accessible to the estimator.

34. _____ Considered to be a good basic experience background for learning to be an estimator.

35. _____ Cost data is added to the total estimate by the different responsible activities.

36. _____ Used to determine processing times.

37. _____ An obsolete type of data storage.

38. _____ Are as harmful to successful estimating as to any other function.

UNIT ▊▊▊

COST ESTIMATING CONTROLS

The purpose of an estimating department is to create realistic cost estimates and other data that company management can use for guidance in decision making for future operations. Therefore, an estimating function is usually organized as a service group for other activities within the organization. To meet all company objectives, an estimating department frequently serves several departments for different purposes and with varying degrees of value to the company's operations. Because cost value and data outputs are related to so many company activities, it is the responsibility of administrative management to dictate the type of estimating organization to be used and to define the company's administrative controls for the use of estimating services and information.

The operating controls used internally by the estimating department should be established by the management of the estimating department and approved by the administrative management. The estimator supervisor must know the limitations and capabilities of his operations and he must understand the requirements of the departments they serve. The cost of operating an estimating department is too great to permit poor workload planning or to allow estimators to spend time on meaningless cost studies. With a clear understanding of the company's objectives and a knowledge of the accuracy of the cost estimates created, a manager of estimating operations can always keep the estimating group's efforts directed toward the company's principal objectives.

Many techniques are used to control the time and effort of an estimator while he is generating cost values. Control of the cost value output data is equally important, for it must be precisely understood by all concerned in order to be useful to the organization. The following are characteristics that are common to all estimating control techniques and output reports:

1. Consistency of format for the publishing date
2. Clear relation of costs to the subject product
3. Continuity and identification of base values used

147

4. Clarity of the product description as estimated
5. Description of manufacturing methods and facilities considered as a basis for processing.

There are pitfalls associated with the establishment of cost estimating controls. Effectiveness can be lost in a maze of administrative procedures intended to control or direct estimating work. Estimators can be buried in paperwork caused by too many data reporting forms. Transposing data from one form to another also increases the probability of clerical errors. Creative effort cannot be identified with realistic or meaningful progress reports and should not be measured by the amount of paperwork incidental to the estimating job.

There is one hazard that company administrative management must guard against when establishing an estimating department. Although the estimator may not make direct decisions affecting company plans or operations, the data which he creates will have a profound effect upon such major decisions. There is a possibility that the potential power of decision available to an estimator can influence his judgment and bias the cost values he creates. The results could give him the opportunity to control major decisions of company management indirectly. One countermeasure to this hazard is to require the estimating department to issue reports to certain responsible areas for review and concurrence before estimates are finally accepted as management information and decision data. A standard, brief summary report, plus adequate detailed information to explain the important features of the estimate, will generally suffice for review by other departments. An estimator must be willing to accept the responsibility of explaining in detail the steps he took in developing his estimate, and he must be able to present his reasons for the results.

Definitions. Briefly define the following:

1. Product mix. _____

_____

2. Marketing department. _____

_____

3. Methods engineering. _____

_____

4. Approving authority. _____

_____

5. Raw material. _____

_____

6. Biased errors. _____

_____

7. Cost indexes. _____

_____

8. Random errors. _____

_____

9. Process engineering. _____

_____

10. Project simplification. _____

_____

Short Answer Questions

11. What advantages are gained by having a screening procedure for estimate requests? _____

_____

_____

12. What is the reason for developing estimates and quotes on all new items?

_____

_____

13. What hazard is greatest in bidding on new or strange items?

_____

_____

14. Name five departments that an estimator usually works closely with.

_____

_____

15. Give two reasons why marketing departments request cost estimates.

_____

_____

16. Describe three methods of estimating. _____

_____

_____

17. List six basic steps which must be performed to prepare a detailed or final design type of estimate. _____

_____

_____

18. Specify what direct control measures can be used to improve estimating accuracy. _____

_____

_____

19. Why does an actual cost deviate from the estimated cost? _____

_____

_____

20. What principles must an estimator understand in order to recognize cost trend changes, and why? _____

_____

_____

True or False

21. An estimator should always question the validity of a request for an estimate.
    True _____ False _____

22. An estimator must maintain rapport with persons in other departments.
    True _____ False _____

23. The estimator himself establishes certain controls to increase the accuracy of his estimates. True _____ False _____

24. An estimating department usually requires a well-planned filing system. True _____ False _____

25. In a process industry most requests for cost estimates come from the marketing department. True _____ False _____

26. An estimator can use his own historical data file to establish a cost index base value. True _____ False _____

27. Cost reduction functions rely upon the estimating department to approve cost reduction proposals. True _____ False _____

28. In preparing estimates for future production, anticipated volume and facilities must be studied. True _____ False _____

29. An estimator can determine and predict his own estimating bias. True _____ False _____

30. Most estimates show relatively large deviations which is a general characteristic of good estimating. True _____ False _____

Matching. Select the appropriate matching word or phrase and record the letter for the word or phrase in the space to the left of each of the following statements.

(a) Detailed analyses
(b) Biased errors
(c) Small deviations
(d) High estimates
(e) Administrative controls

(f) Comparison method
(g) Low estimates
(h) Forms
(i) Random errors
(j) Conference method

31. _____ Can be controlled where sufficient historical data have been accumulated.

32. _____ Follow trends that may be due to assignable causes.

33. _____ Include monitoring of incoming cost requests and the establishment of routings for completed estimates.

34. _____ May be used to pool the specialized knowledge of expert estimators.

35. _____ Are a general characteristic of good estimating practice.

36. _____ Are among the best devices to control estimating methods for continuity, consistency, and completeness.

37. _____ Represent the hazard of a potential loss to the company.

38. _____ Relies upon past experience and data updated for latest cost values.

39. _____ Cause rejection of proposals that could be profitable.

40. _____ Are the most reliable tools for furnishing the most accurate prediction of anticipated costs.

UNIT IV

ESTIMATING PROCEDURES

Complexities created by the involvement of an estimating department in the routine operation of a manufacturing concern require that definite operating procedures be established for the department. Operating procedures are required as guidelines to ensure that the flow of output information will meet the requirements of the established cost control system. However, the procedures must not be allowed to hamper the freedom of an estimator in using any technique he may choose as a method of creating the objective estimated cost values. The procedures for estimating as outlined in Chapter 4 encompass basic techniques that are logical approaches to developing an estimated cost, regardless of what the product might be.

In the first four chapters of the text, care is taken to show how the estimator works within the complex activities of a company operation. Yet the principal output bits of information generated by an estimator are ultimately refined and summarized in four primary categories of manufacturing cost data. They are as follows:

1. Direct material costs which represent the delivered cost of the materials that enter directly into the finished product
2. Direct labor costs that represent wages of the workmen who convert the materials into a finished product
3. Factory overhead or burden costs which include all of the remaining costs of operating a factory
4. Costs for tooling and/or additional facilities that will be required for production of the proposed product.

An estimator must thoroughly understand the financial analysis aspects of each of these cost categories and the types of decisions to which the cost information will be applied. In Chapter 1 a brief explanation is given for the various types of manufacturing costs and the cost-volume-profit relationships. The content of Chapter 4 expands on these definitions of costs and their applications to the cost accounting categories. Because most cost estimators have come up from the ranks of production or manufacturing engineering departments—and retain knowledge and skills associated with these departments—their background and experience have not included the increased orientation toward finance now being demanded of cost estimators.

To assist in developing broader understanding of the procedures and financial analysis aspects of cost estimating, Unit IV is divided into five parts to deal with each estimating procedure or cost category individually. The student estimator is encouraged to research each cost area on his own more deeply than it is covered in this book. Academic study of the basic principles of cost accounting and management accounting are a worthwhile starting point. Publications of the technical societies and trade associations related to specific industries or technologies are excellent sources of financial information to assist an estimator. Finally, establishment of close working relationships with personnel in the financial offices of the student's own organization has obvious advantages for all concerned.

UNIT IV—PART A

Procedural Approach to Developing a Cost Estimate

The seven basic steps and the checklist of items outlined on pages 35 and 36 in the textbook illustrate the logical sequence of events which occurs in the development of any estimated cost. The estimating procedure used ultimately generates the needed data at a depth of detail or accuracy classification as shown in Table IV-1, page 37. The class of estimate generated is contingent upon the time limit and depth required for reasonable accuracy, using the product and manufacturing data available.

As mentioned on page 36, an estimator must have the imagination to visualize his future needs for technical knowledge and data about new material or processing technologies. And he must take the initiative to systematically accumulate such knowledge in anticipation of a need for its use.

Definitions. Briefly define the following:

1. Standard parts. _____

_____

2. Standard time data. _____

_____

3. Manufacturing routing. _____

_____

4. Bill of materials. _____

_____

5. Plant facilities. _____

_____

6. Millwright. _____

_____

7. Due date. _____

_____

8. Processing plan. _____

_____

9. Profit margin. _____

_____

10. Fabricated part. _____

_____

Short Answer Questions

11. What items are usually not specified on layout drawings? _____

_____

_____

12. Give two reasons for subcontracting work. _____

_____

_____

13. What is helpful when using a comparison method for estimating a cost?

_____

_____

_____

14. Briefly describe the two extremes in modes of estimating operations. How are they different and what do they have in common? _____

_____

_____

15. What preventive measure does an experienced estimator take?

_____

_____

16. Name three items that would be included in a facilities cost estimate.

_____

_____

17. List four factors to be analyzed in estimating costs for new facilities.

_____

_____

18. What three variables determine the class of estimate shown in Table IV-1?

_____

_____

19. Is Parkinson's law realistic or facetious? State your opinion. _____

_____

_____

20. Why does an efficient operation require formalized estimating procedures and historical records? _____

_____

_____

UNIT IV—PART B

Direct Material Cost

A definition of *direct material cost*, also called *raw material cost* or *stores*, is the delivered cost of the material that enters directly into the finished product. This cost normally includes the inbound transportation, handling, and storage costs associated with delivery of the material to the customer's manufacturing plant. The amount of direct material used to manufacture an item includes the material used to make the single unit plus prorated allowances for scrap losses that are inherent in the manufacturing process. The quantity of material used per unit multiplied by the unit cost of the material equals the direct material cost per unit.

The examples given in the text on pages 41 and 42 are typical for steel bar stock and sheet steel material estimating procedures. These same principles would apply if the material were a new exotic metal alloy, plastic, paper, or wood. The important point is that an estimator must recognize the correct approach to the particular material technology used and be able to visualize the allowances to be expected with various manufacturing processes.

Another aspect of estimating material costs is in understanding and interpreting the product engineering materials specifications and their relationship to the choice of manufacturing process. Some product engineering materials specifications allow latitude in the choice of the material which can be used to make the end product. This flexibility gives the cost estimator a range of choices in materials and processes to obtain the material/process combination with the lowest cost. One example is the choice between the use of hot-rolled or cold-rolled steel. The hot-rolled steel is normally lower in cost, but if the cost advantage is lost because of extra labor or equipment needed to finish its inherently rougher surface, no cost savings can be realized. The cost estimating procedure is a series of judgments in choosing the best combination of material, process, and equipment.

Definitions. Briefly define the following:

1. Material density. _____

_____

2. Material specifications. _____

_____

3. Bar stock. _____

_____

4. Standard part cost. _____

_____

5. Offal. _____

_____

6. Butt ends. _____

_____

7. Standard cost (materials). _____

_____

8. In-house fabrication. _____

_____

9. Stock machining allowance. _____

_____

10. Coil stock. _____

_____

Short Answer Questions

11. Why should precautions be taken when cost tables for standard purchased parts are used? _____

_____

_____

12. What is a slitting or shearing operation? _____

_____

_____

13. Why do some companies have a slitting or shearing operation of their own? _____

_____

_____

14. Where is the cost for slitting or shearing included in a cost estimate?

_____

_____

15. Why would a material purchase price break be applied to an individual product part cost? _____

_____

_____

16. Where are standard costs most useful? _____

_____

_____

17. Identify the two basic categories of material. _____

_____

_____

18. What is the hazard of estimating too much stock machining allowance?

_____

_____

19. What is the danger of estimating too little stock machining allowance?

_____

_____

20. Who establishes standard costs, and how are they computed?

_____

_____

UNIT IV—PART C

Direct Labor Cost

Direct labor costs consist of wages and other labor-related costs incurred in the payment of production workers who are engaged directly in specific manufacturing operations to convert direct materials into finished products. Direct labor costs can either be identified specifically with a unit of the product or have a consistent value which is closely related to a given production quantity so that a direct cost relationship can be established. The direct labor workers are those who perform the repetitive processing tasks which change the characteristics of direct materials—those who operate production machines or processing equipment, assemble parts into a finished product, or work on the product with tools. All other manufacturing labor costs (salaries for supervisors, tool and die makers, repairmen, storekeepers, custodians, and other plant operating personnel, for example) are called indirect labor costs and are included as part of manufacturing overhead costs.

One direct labor cost factor commonly used by manufacturing companies for practical cost estimating purposes is the basic hourly wage rate plus an hourly rate or percentage of wages for the fringe benefits credited to each worker's personal account. A second factor is the estimated average production quantity that can be realistically attained by a worker in a specified period of time—normally stated as a quantity per hour. By using rate per hour as the common denominator, an estimated direct labor cost per unit is obtained as follows:

$$\frac{\text{Wages and benefits in dollars per hour}}{\text{Estimated production attainable per hour}} = \text{Estimated direct labor per unit}$$

Most industries with high-volume production rates and long-term production runs identify the estimated labor effort in terms of time study data, and they express the labor per unit by values of standard minutes per piece. The use of time standards simplifies the computation for many different manufacturing planning purposes. To obtain a unit cost in terms of standard minutes per piece, the cost estimator first converts the hourly wage rate to a per minute wage rate and then multiplies by the standard minute value. Another technique uses a time study value based on a decimal equivalent per hour which is multiplied by the hourly cost rate. Some estimators use an estimate of labor time per thousand units and multiply by the equivalent labor cost rate to obtain a cost per thousand value.

Many techniques can be used, but most modern managements are using techniques that are, or will be, compatible with data processing input formats. Some companies use a straight piece rate cost system which gives a precise estimated labor cost value per unit of production. Others have various base values plus an incentive bonus plan for direct labor costs. Regardless of the factoring method used, all direct labor costs are ultimately expressed in a format that is meaningful to the financial and operating management.

A cost estimator must recognize the importance of clearly identifying the assumptions he uses in developing direct labor costs. He should always label the type and amount of labor rate that was used. He must show the operating time system used, such as one or two shift operation with straight time or a scheduled overtime planned for a specified period. The estimator should indicate if the fringe benefit costs are included or shown separately. Finally, he should always be consistent in his format of expression or else identify what is different from conventional practices previously used.

Confusion is sometimes generated because of differences in the denotation of certain financial terms and the connotation intended by the estimator. The term *cost* is a typical example. From an accounting standpoint, the term *cost* is generally used to identify those costs which go into the finished product and are identifiable as part of the product. The word *expense* is usually applied to those items which are part of the cost of doing business but which cannot be specifically related to the end product. The experienced estimator will determine what the conventional uses of such terms are within his organization.

Prime cost is a term used for a specific type of management decision factor for planning. Prime cost is the sum of the direct material costs and the direct labor costs of a product. Prime costs may be used in decisions for making minor product design changes or minor manufacturing process changes that would not produce a meaningful effect on the overhead operating costs.

To predict direct labor costs the estimator must have a thorough understanding of basic time-and-motion study data and a clear knowledge of the meanings of the accounting terms used to portray labor cost values.

This unit has purposely skipped over the subject of tooling costs and manufacturing time periods outlined on pages 44–47. The subject of manufacturing process time and its relationship to both direct material and direct labor will be covered by the next unit.

Definitions. Briefly define the following:

1. Hourly wage rate. _____

_____

2. Fringe benefits. _____

_____

3. Production time. _____

_____

4. Time study. _____

_____

5. Prime cost. _____

_____

6. Direct labor. _____

_____

7. Indirect labor. _____

_____

8. Repetitive processing tasks. _____

_____

9. Time standards. _____

_____

10. Standard minutes. _____

Short Answer Questions

11. Which of the following job classifications would logically be considered as direct labor? Circle the letters identifying the correct answers.

   a. Punch press operator

   b. Assembler (final line)

   c. Tool crib attendant

   d. Automation machine operator

   e. Job setter (machines)

   f. Sewing machine operator (upholstery)

   g. Spray painter (assembly line)

   h. Shipping clerk

   i. Polisher (plating department)

   j. Floor inspector (quality control)

12. Which of the following items would normally be included in the fringe benefits cost factor? Circle the letter identifying the correct answers.

   a. Cost of living allowance

   b. Vacation pay

   c. Scrap cost factor

   d. Pension plan costs

   e. Hourly shift premium

   f. Employee parking lot maintenance cost

   g. Holiday pay

   h. Union steward's wages

   i. Company paid hospitalization insurance

   j. Social Security costs, company contribution

13. The base hourly wage rate for a direct labor job is $3.20 per hour, and the fringe benefits cost to be added is 25 percent of the base rate.

   a. What is the net direct labor cost rate per hour?

   b. What is the net direct labor cost rate per minute?

14. Time-and-motion study data indicates a net practical production of 600 units per hour. What are the values for the following factors?

   a. _____ seconds per piece

   b. _____ pieces per minute

   c. _____ time standard minutes per piece (express in decimal minutes to fourth place)

   d. _____ time standard hours per piece (express in decimal hours to fifth place)

15. Develop a direct labor cost per piece by using the following formulas and

data, 600 units per hour, $3.20 per hour base rate includes 25 percent fringe benefit rate.

a. $\dfrac{\text{Net hourly wage}}{\text{Units per hour}} =$ _____

b. Standard minutes per piece × net labor cost per minute = _____

c. Standard hours per piece × net labor cost per hour = _____

d. $\dfrac{\text{Net wage rate per minute}}{\text{Net practical per minute production}} =$ _____

16. What is the prime cost based on the following data? _____

Direct material cost is $0.0100 per piece.

Direct labor cost is that developed in Question 15.

UNIT IV—PART D

Factory Burden

The term *factory burden* is one of the more popular terms used to describe manufacturing costs that cannot be specifically attributed to or equated with the product. Other terms in use are *manufacturing overhead, factory overhead, operating burden,* and *overhead costs.* The cost of factory burden is usually allocated to the product by applying a predetermined burden rate factor. The burden rate (also known as absorption rate) is the percentage of applicable manufacturing expense to the standard direct labor cost at control volume.

Burden rates are developed by the accounting office function and then are given to the estimator for use. An estimator must understand the elements used to develop a burden rate. If the process/method planned in a cost estimate study is capable of significantly changing the precedent burden rate value that is being used, the estimator must bring the matter to the attention of responsible persons for their consideration. The wide variations in burden rates associated with modern manufacturing processes which use automated or sophisticated equipment can be a major factor in making business decisions.

Identify the following items as fixed or variable by the category of burden they represent.

	Fixed	Variable
1. Land taxes	_____	_____
2. Perishable tools	_____	_____

3. Inspection labor   _____   _____

4. Insurance premiums   _____   _____

5. Equipment maintenance   _____   _____

6. Utility charges   _____   _____

7. Spoilage of material   _____   _____

8. Fuel for heat treatment furnaces   _____   _____

9. Supervision salaries—administrative _____   _____

10. Wages—material handlers   _____   _____

Short Answer Questions

11. Name three common base factors used in assigning burden rates.

_____ , _____ , _____ .

12. What type of industries would most likely use the direct material cost as a factor for burden costs? _____

_____

_____

13. What is a cost center? _____

_____

_____

14. Describe what affects variable burden costs. _____

_____

_____

15. Define fixed and variable burdens.

Fixed burden. _____

_____

Variable burden. _____

_____

UNIT IV—PART E

Manufacturing Time and Methods

Determining optimum manufacturing methods and outlining the sequence of process operations and the equipment to be used are two of the estimator's most creative tasks. From these determinations the estimator develops the material quantities, the direct labor times, and the burden rate applications required. To these elements he applies the cost rate factors as previously covered in Parts B, C, and D of this unit. He must also consider cost values developed for the tooling and facilities that will be required to set up and initiate the production run.

The modern estimator must be capable of performing all the creative tasks that historically were performed by various persons with many different specialized skills. The urgency of management decisions, or the small size of some organizations, will frequently not allow him time to research, compile, and summarize the data from many different specialists. Veteran estimators have acquired the basic knowledge of process, production, tool, method, plant layout, industrial, and time-study engineering. Their creativity is attained by their employment of an appropriate degree of understanding of all the required skills in order to visualize and place logical values on the methods that will produce desired products for the most economical costs.

The best production method is generally found by performing a series of exploratory computations and analyses based upon historical data, testing the various alternative results, and selecting the best one. Each step in this elimination process will become more detailed as the search is made for the one best way. The final selection will be the method or process which is the most economical and practical for the production operation concerned.

In some manufacturing organizations estimators develop the method/process cost rationale for new products which actually are developmental modifications of current specialty products. The proposed new product designs are tailored to suit most of the facilities and personnel available at that time within the company. A cost for the new product is estimated by using the actual cost data factors available for the current product version and by judging new values for changes to modified current product costs or processes.

Other types of manufacturing organizations specialize in a specific method or process (small stamping, diecasting, gear cutting, plating, heat treatment, etc.) and seek customer orders for products which are suitable for manufacturing with their usual facilities and personnel. Good record maintenance results in the accumulation of a bank of historical case study data to provide adequate reference information in such cases. Adoption of current production data to estimate costs is a realistic technique, for careful modifications of existing values are just as accurate as the new values obtained by induction.

The most difficult task confronting the estimator is that of estimating costs for a new technology. New products made from new kinds of materials and obviously needing a new technological approach to manufacturing, provide little empirical data to use for base values. The estimator is first challenged to develop his technical knowledge about the new product and/or new material. Next, he must envision a new method and sequence of processing operations to attain the desired product and then assign values to each material or manufacturing operation. Finally, he must develop values for the tools and facilities that will be required to make the product.

New technologies demand creativity in cost estimating. Estimating the cost of a product to be manufactured with new technology is a good test of the estimator's ability to solve problems through imagination and judgment.

Terms. Match the following related terms for their best associations. Some terms have more than one association. Place the letter indicating your choice in the left-hand column.

1. _____	Index and idle time	a. Iron foundry
2. _____	Bend radii	b. Diecasting
3. _____	Cope and drag	c. Automatic machining
4. _____	Case hardening	d. Sheet metal stamping
5. _____	Feed and speed	e. Standard time data
6. _____	Part orientation	f. Gear cutting
7. _____	Shot	g. Painting booth
8. _____	Inherent delay	h. Automatic assembly
9. _____	Press fit	i. Heat treatment
10. _____	Anneal	j. Machinability rating
11. _____	Ladle time	k. Blank diameter
12. _____	Shearing strength	l. Product dimensions
13. _____	Machine horsepower requirement	

14. _____ Draw quality

15. _____ Stripper

16. _____ Hobbing

17. _____ Personal allowances

18. _____ Soak time

19. _____ Overspray allowances

20. _____ Pitch diameter

21. _____ Time and temperature

22. _____ Offstand variance

23. _____ Patterns

24. _____ Strokes per minute

25. _____ Stack of tolerances

ANSWERS

UNIT I

Definitions

1. Process whereby the total manufacturing cost of a product is estimated.

2. Estimated cost of an entire product as opposed to the estimated cost of its individual components.

3. An individual component, generally one of a number of components that make up an assembly.

4. A subsystem comprising two or more components.

5. Costs that are directly related to a specific piece part.

6. The cost of labor directly associated with the manufacture of a product.

7. Necessary costs (e.g., janitorial, maintenance, lighting) not directly associated with any specific piece part or product.

8. Necessary labor costs not directly associated with any specific piece part or product.

9. Composite cost (generally an hourly figure) comprising all indirect costs incidental to plant operation.

10. Costs necessary to plant operation but not associated with production. Executive salaries, R & D, and public relations are examples.

11. An ideal cost—the amount a product *should* cost barring unforeseen contingencies.

12. The cost of a product—established by the cost accounting department after the product has been manufactured.

13. Unchanging costs that are independent of the production volume. Executive and secretarial salaries are examples.

14. Direct costs that vary as a function of the production volume. Labor costs and material costs are examples.

15. Costs that remain constant over a given number of parts and then increase sharply as additional production equipment is added.

16. The process of determining the actual cost of a part.

17. The process of reducing the cost of manufacturing a part.

18. The process of optimizing product design through systematic analysis of all design alternatives.

19. The process of optimizing all aspects of the manufacturing process.

20. Product analysis to determine which components shall be manufactured in-house and which shall be purchased.

21. Cost estimate made — for management benefit — prior to finalization of product specifications.

22. Cost estimate made *after* all specifications are finalized and *before* production begins.

23. Preliminary specifications — used in the absence of part drawings — to establish preliminary estimates.

24. Part of the preliminary cost estimate constituting the best possible approximation of the cost of tooling.

25. Estimated cost of facilities needed.

Short Answer Questions

26. Management

27. None

28. Cost estimating is for predicting costs whereas other financial functions deal with past events.

29. Because no manufacturing plan can be initiated without it.

30. How closely is the cost related to part manufacture? How practicable is it to relate cost to a certain product?

31. When a relevant range can be defined.

32. (1) Feasibility of part manufacture, (2) In making make-or-buy decisions,

(3) In bidding on contracts, (4) In evaluating the products of competitors or vendors.

33. In evaluating alternative product designs.

34. Up-to-date knowledge of production processes, tooling and materials.

35. That of a performance standard.

True or False

36. False

37. True

38. True

39. True

40. False

41. False

42. True

43. False

44. False

45. True

46. True

47. True

48. False

49. True

50. True

51. True

52. True

53. True

54. False

55. True

Matching

56. g

57. c

58. f

59. h

60. j

61. b

62. a

63. e

64. i

65. d

ANSWERS

UNIT II

Definitions

1. Responsibilities for certain activities or functions that are assigned to a given department or individual.

2. Associated with difficulties encountered in product manufacture.

3. The manufacturing engineering function dealing with the development of work standards through optimization of operator movements.

4. Reduced prices offered as part of a sales strategy.

5. A manufacturing concern which has no product of its own but supplies parts or tooling to customer specifications. Job shops are often called "contract" shops.

6. Rough guesses as to eventual price.

Short Answer Questions

7. An analytical mind. Efficiency. Initiative.

8. A bid estimate consists of the cost estimate plus a margin of profit. A cost estimate is an estimate of the cost to manufacture a given product.

9. Manufacturing engineering. Purchasing. Process planning. Planning and scheduling. Plant layout.

10. Integrated and departmental. In an integrated system a single department does the estimating. In a departmental system all departments associated with the projected product furnish cost estimates concerning their portions of the total manufacturing operation.

11. An analytical approach. Knowledge of time-and-motion study and of the manufacturing processes.

12. Because the sales function introduces a powerful bias in favor of sales at the expense of profit.

13. Added flexibility in that a few estimators can serve a relatively broad product line.

14. Greater efficiency in the development of estimated costs.

15. Faster estimate preparation. Greater accuracy. More economical utilization of manufacturing equipment. More efficient use of estimating personnel.

16. As a storage facility for data acquired through experience. In making high speed calculations. In some instances in preparing the total estimate.

True or False

17. True

18. False

19. True

20. False

21. False

22. True

23. False

24. True

25. True

26. False

27. False

28. True

Matching

29. b

30. a

31. d

32. f

33. e

34. c

35. i

36. g

37. j

38. h

ANSWERS

UNIT III

Definitions

1. The variety of products produced by a company.

2. The department that is assigned responsibility for all aspects of merchandizing the company's products.

3. The department that analyzes a proposed product to determine manufacturing time requirements.

4. The department or executive that gives final approval to the cost estimate.

5. The basic materials from which the finished product will be made.

6. Trend type errors due to assignable causes, i.e., fluctuations in labor and material costs, shortages in vendor capacity, etc.

7. Published charts in which cost of products in a given year is based on the cost of each product in a certain, arbitrarily selected, base year.

8. Chance errors, normally assignable to specific causes which can be detected and controlled.

9. The systematic development of a sequence of manufacturing steps.

10. Reducing a project or product to its simplest components, the objective being to minimize the possibility of estimating errors.

Short Answer Questions

11. The screening procedure eliminates all requests for bids on products not desired by the company.

12. To establish approximate cost of manufacturing and thus to determine manufacturing feasibility.

13. The absence of a data base.

14. Purchasing, Process Planning, Methods Engineering, Plant Engineering, and Accounting.

15. (1) To fulfill a customer request. (2) To determine manufacturing feasibility.

16. Conference—in which representatives of all departments concerned provide estimates concerning their portion of the work. Comparison—in which estimator bases cost on the cost of similar products produced in the past. Detailed Analysis—in which estimator makes a detailed breakdown of the job, and estimates the cost of each phase of the manufacturing requirements.

17. (1) Calculate material requirements, (2) process the individual components, (3) compute direct labor cost for each step, (4) determine equipment required, (5) determine tooling required, and (6) evaluate inspection requirements.

18. The use of machine and worker performance factors, estimate simplification, and cost indexes.

19. Because of human error, inadequate estimating procedures, and the fact that not all factors affecting cost can be recognized.

20. The principles behind the cost accounting system.

True or False

21. False

22. True

23. True

24. True

25. False

26. True

27. True

28. True

29. True

30. False

Matching

31. i

32. b

33. e

34. j

35. c

36. a

37. g

38. f

39. d

40. h

ANSWERS

UNIT IV—PART A

Definitions

1. Items such as screws, dowel pins, and handles that are produced to standardized dimensions and are obtainable through vendors.

2. Data concerning time requirements for standardized operations. Use of these data obviate the need to calculate time requirements for certain operations.

3. A list of the various points in the facility through which a part must pass in the manufacturing process.

4. An itemized list of all materials and standardized parts required for a product.

5. Capital equipment required for manufacturing.

6. Employee assigned responsibility for setup and maintenance of plant facilities.

7. The date the finished parts—for which an estimate is being prepared—must be delivered.

8. The logical sequence of manufacturing operation required to produce a part.

9. The differential between part sales price and total cost of producing the part.

10. Part that must be manufactured as opposed to a part that can be purchased.

Short Answer Questions

11. Detail dimensions, tolerances, processes to be used and materials.

12. (1) Greater profitability in doing so, (2) Because plant facilities are overloaded.

13. A good filing system containing previous estimates for similar parts.

14. Extremely simplified vs. highly detailed. In the former the estimator performs largely on the basis of experience and intuition. In the latter he bases his estimate on a detailed breakdown of the product, analyzing each element for cost. The common ground is a data file—in a filing cabinet or locked in memory.

15. He develops a file of data for use in making comparison estimates.

16. Rearrangement of machines, relocation of utility outlets, and alteration to the building structure.

17. (1) Area where facilities are to be placed, (2) existing facilities, (3) required facilities, (4) new product flow pattern.

18. (1) Information needed, (2) method of estimating, and (3) time available for estimating.

19. A real condition described in a facetious manner.

20. Because formalized procedures and historical records tend to reduce the range of error.

ANSWERS

UNIT IV—PART B

Definitions

1. The ratio of its mass to its volume.

2. Type, form, and dimensions of the material order. Specifications may also include other criteria such as hardness and finish.

3. Bars of material—square, round, or hexagonal—which are fed to lathes, screw machines, bar automatics, etc., to produce machined parts.

4. The cost of various standard components (screws, dowels, etc.) usually kept on file or in computer memory to expedite estimating.

5. Scrap sheet metal from stamping operations, often used in the production of other, smaller, stampings.

6. Ends of bar stock. Because these must be "cleaned up" they represent scrap loss.

7. A pre-established per-part material cost. To determine total material cost, the estimator simply multiplies standard cost by number of parts.

8. The in-house manufacture of certain components as to the purchase of these components from vendors.

9. The amount of stock to be machined away. Thus a 0.9375 diam. part turned from 1-inch stock has a 1/16 inch machining allowance.

10. Coiled sheet metal for stamping operations.

Short Answer Questions

11. Because prices vary frequently.

12. The cutoff by machine of sheet metal from coiled stock. Flat sections thus obtained are then fed to the press.

13. They perceive certain economics in preparing flat stock for subsequent press operations.

14. In the material costs.

15. To obtain an extremely precise per-part cost figure.

16. In job shops where a wide variety of parts are produced.

17. Sheet stock and bar stock.

18. Excessive waste of raw material and machining time.

19. That the part may not "clean up" and—in the case of tool steel—that the decarburized layer may not be removed.

20. The accounting department. By calculating the average cost of material per-part over an extended period of time.

ANSWERS

UNIT IV—PART C

Definitions

1. The per/hour wage earned by an employee.

2. Employee benefits in the form of paid vacations, hospitalization insurance, etc., that must be added to the wage rate to obtain a net hourly rate.

3. The total time spent in producing a part.

4. The process of determining the precise amount of time required to perform a single operation.

5. The sum of the direct labor costs plus the direct material costs.

6. The labor required to produce a part.

7. The labor required to run a business but not directly associated with specific part production.

8. The repetitive tasks normally associated with production, e.g., running a punch press, drilling holes, operating a spot welder.

9. The intervals of time—established through time study—required to perform the various manufacturing operations.

10. The measurement of labor required to perform various manufacturing operations.

Short Answer Questions

11. a, b, d, f, g, i.

12. a, b, d, e, f, g, i, j.

13. a. $\underline{\$4.00}$

  b. $\underline{\$0.066}$

14. a. $\underline{6}$

  b. $\underline{10}$

  c. $\underline{0.1000}$

  d. $\underline{0.00166}$

15. a. $\dfrac{\$4.00}{600} = \$0.00666$

  b. $\underline{\$0.0066}$

  c. $\underline{\$0.0066}$

16. d. $0.0166

ANSWERS

UNIT IV – PART D

1. Fixed

2. Variable

3. Variable

4. Fixed

5. Variable

6. Variable

7. Variable

8. Variable

9. Fixed

10. Variable

Short Answer Questions

11. (1) Direct labor cost, (2) Direct material cost, (3) Number of parts produced.

12. Process type industries

13. A department or division assigned its own budget and its own burden rate.

14. Variations in the production volume.

15. Fixed burden: Expenses associated with the cost of doing business, i.e., taxes, administrative salaries, insurance, and the like.

    Variable burden: Costs that vary as a function of production volume, i.e., materials electrical power, and so on.

ANSWERS

UNIT IV—PART E

Terms

1. e

2. d or l

3. a

4. i

5. j

6. l

7. a or i

8. e

9. l

10. i

11. b or a

12. j or l

13. j

14. d

15. d

16. f

17. e

18. i

19. g

20. f

21. i

22. e

23. a

24. d

25. h